Ecoambiguity, Community, and Development

Ecocritical Theory and Practice
Series Editor: Douglas Vakoch, California Institute of Integral Studies, USA

Ecocritical Theory and Practice highlights innovative scholarship at the interface of literary/cultural studies and the environment, seeking to foster an ongoing dialogue between academics and environmental activists. Works that explore environmental issues through literatures, oral traditions, and cultural/media practices around the world are welcome. The series features books by established ecocritics that examine the intersection of theory and practice, including both monographs and edited volumes. Proposals are invited in the range of topics covered by ecocriticism, including but not limited to works informed by cross-cultural and transnational approaches; postcolonial studies; ecofeminism; ecospirituality, ecotheology, and religious studies; film/media and visual cultural studies; environmental aesthetics and arts; ecopoetics; and animal studies.

Recent Titles

Transversal Ecocritical Praxis: Theoretical Arguments, Literary Analysis, and Cultural Critique, by Patrick D. Murphy

Ecoambiguity, Community, and Development: Toward a Politicized Ecocriticism, edited by Scott Slovic, Swarnalatha Rangarajan, and Vidya Sarveswaran

Ecoambiguity, Community, and Development

Toward a Politicized Ecocriticism

Edited by Scott Slovic,
Swarnalatha Rangarajan,
and Vidya Sarveswaran

LEXINGTON BOOKS
Lanham • Boulder • New York • Toronto • Plymouth, UK

Published by Lexington Books
A wholly owned subsidiary of Rowman & Littlefield
4501 Forbes Boulevard, Suite 200, Lanham, Maryland 20706
www.rowman.com

10 Thornbury Road, Plymouth PL6 7PP, United Kingdom

British Library Cataloguing in Publication Information Available

Library of Congress Cataloging-in-Publication Data
Ecoambiguity, Community, and Development Toward a Politicized Ecocriticism / edited by Scott Slovic, Swarnalatha Rangarajan and Vidya Sarveswaran.
pages cm. — (Ecocritical Theory and Practice)
ISBN 978-0-7391-8908-5 (cloth : alk. paper) — ISBN 978-0-7391-8909-2 (ebook)
ISBN 978-1-4985-2536-7 (pbk : alk. paper)
1. Ecocriticism. I. Slovic, Scott, 1960- editor of compilation. II. Swarnalatha, R., editor of compilation. III. Sarveswaran, Vidya, editor of compilation.
PN98.E36E27 2014
809'.9336—dc23
2013049002

∞™ The paper used in this publication meets the minimum requirements of American National Standard for Information Sciences Permanence of Paper for Printed Library Materials, ANSI/NISO Z39.48-1992.

Printed in the United States of America

Contents

Introduction

Scott Slovic, Swarnalatha Rangarajan, and Vidya Sarveswaran

The opening example of "ecoambiguity" in Karen Thornber's 2012 book *Ecoambiguity: Environmental Crises and East Asian Literatures* describes a situation in contemporary Japan where a local tourism association at Shosen Gorge sought to cut down trees "so that people could have a better view of 'nature'" (1). In the case of Shosen Gorge, an economically depressed region that relies on income from tourists, visitors are particularly keen to have unobstructed views of the spectacular rocky cliffs, so the natural vegetation has been targeted for cutting in order to facilitate an aesthetic or touristic experience. Essentially what Thornber is describing is a cultural tendency to selectively appreciate and resist the natural world, seeking to control our experience and the natural environment as a way of maximizing a desired mode of experience. In many other parts of Asia and the rest of the world, the irony of ecoambiguity is more painful and acute. This collection extends the work of Thornber's important study and locates a wide range of paradoxical and destructive ecoambiguities in disparate regions of the world, from China to South Africa. What is unique in this project is that the many of the contributors to this work come from the places, or have been associated with these places, they are writing about—there is an intimacy and a ferocity in these efforts to "politicize" ecocritical inquiry that may be less evident in many other works in the field. At the same time, some of the leading First World theorists of environmental culture in the developing world, from South Asia to Latin America, have offered scholarly overviews that will give contextual anchors for the more regionally specific studies.

At first glance, this collection may seem to contain a rather diverse selection of articles. What holds the pieces together, though, are the themes of community and development, generally seen through a somewhat ironic lens—that of "ecoambiguity." Thornber concludes her 2012 book by arguing that "Ecocriticism and other branches of environmental humanities demonstrate especially clearly the exciting possibilities for humanistic intervention in ecodegradation" and looking forward to collaborations that will enable scholars across the disciplines to "embrac[e] more fully the wider world—culturally, geographically, biophysically"

(435). This new book seeks to do precisely that—to reach out and take stock of cultural and environmental contexts in many different regions of the world by exploring literature and film in these places. One of the complications that quickly becomes clear, as artists and scholars working in the social ecology, environmental justice, and postcolonial arenas have long recognized, is that as soon as we begin to tug on a thread of "ecodegradation," we generally find it linked to some form of cultural oppression. The reverse is also often true. So the studies collected here, in the spirit of postcolonial ecocriticism, emphasize the impossibility of disentangling environmental and cultural problems. Although not all of the authors explicitly invoke the term "ecoambiguity"—or the concepts and terminology of postcolonial ecocriticism, for that matter—their articles frequently bring to light various ironies, such as the fact that Ukrainian environmental experience in the twenty-first century is defined by one of the world's most infamous industrial disasters (the Chernobyl nuclear accident of 1986), and yet Ukrainian culture, like many throughout the world, actually cherishes a profound, even animistic, attachment to the wonders of nature. The repetition of this and other paradoxes in human cultural responses to the more-than-human world reinforces our sense of the congruities and idiosyncrasies of human culture. Every human culture, regardless of its condition of economic and industrial development, has produced its own version of "environmental literature and art"—but the nuances of this work reflect that culture's precise social and geophysical circumstances. In various ways, these stories of community and development from across the planet converge and diverge, as told and explained by distinguished scholars, many of them coming from the cultures represented in these articles.

We begin this collection with Karen Thornber's own article, which builds upon her earlier work on ecoambiguity in East Asian environmental literature. In this new essay, she focuses on the vexed nature-culture interactions on the northern and southern borderlands of China, as explored in Chinese literature. The contradictory human interactions with nature in China are graphically evident in the examples of Chinese writing that both celebrate natural beauty and refer to nature's degradation at the hands of society. In this particular study, Thornber emphasizes two specific modes of ecoambiguity: the rapid fluctuation of attitudes between awe and indifference and the situation (familiar throughout the world) of loving nature so much that we extract too much from it or crush it with our own presence. These and other modes of ecoambiguity are evident throughout Chinese culture, as is seen in Thornber's 2012 book. But for various reasons, the conditions mentioned above are particularly vivid in the ecologically sensitive regions along China's southern border with Laos and Burma and its northern region known as Inner Mongolia. As Thornber explains, "most literature on environmental distress further ambiguates these ambiguities, bringing to light deeper discrepancies." The function of this literature is not necessarily to solve practical problems of ecodegradation and social struggle, but to intensify public and governmental attention to issues that demand conscious efforts at correction. What is unseen and unnoticed is more likely to fester and worsen. Even in a country like China, where there may be limited give and take, restricted communication between the public and official representatives, this rule holds true—if the vast public becomes aware of social and environmental problems, the government is under pressure to respond.

Gang Yue, in his study of ecopolitics in Tibet, another region on the southern border of China, tests this hypothesis further. He begins by discussing the popular contemporary blog called *Invisible Tibet* by Beijing-based Tibetan writer Tsering Woeser, which tracks ecological and social distress caused by Chinese mining in Tibet. The very title of this blog connects powerfully with the textual examples mentioned in Thornber's article, both of which aim, in part, to make the far-away, invisible borderlands somehow more visible to mainstream, urban Chinese citizens. But, as Gang Yue demonstrates, the Tibetan situation has unique features, particularly because of the highly politicized status of Tibet, perhaps the most globally famous example of Chinese territorial disputes. Whether one recognizes Tibet as an autonomous, non-Chinese nation or believes that Tibet is a traditional part of China, the fact that Tibet is disputed territory is quite evident. The link between this political sensitivity and environmental degradation, however, is less well understood and is the focus of the article in this book. While much of this article explores the political situation in Tibet, Gang Yue also addresses the role of literature and film in inspiring both native and international resistance to environmental destruction in this part of the world, focusing on texts as disparate as naturalist George Schaller's *Tibet Wild* (2012) and the 2004 film *Kekexili*. The status of environmental intervention in Tibet is fraught with ambiguity, even potentially calling into question whether drawing further international attention to this politically sensitive region will facilitate cultural and ecological protection or merely inflame Han Chinese efforts to crack down on Tibetan autonomy.

The relationship between mainstream (Han) Chinese culture and the cultural and practical needs of China's ethnic minorities (and the often remote bioregions where these people live—remote from the Beijing-based government in the eastern part of the country) readily calls to mind questions of social and environmental justice. However, the field of Chinese environmental justice ecocriticism is just beginning to gain momentum. Chinese graduate students Cheng Li and Yanjun Liu have produced a strikingly forceful contribution to this topic, using three recent works of Chinese literature to show how authors are vocally engaging with such issues as marginalized Chinese farmers and displaced ethnic minorities, concerns that have become widespread in twenty-first-century China, byproducts of the country's economic success story. By applying vocabulary from environmental justice and postcolonial scholars in other parts of the world to the Chinese context, Li and Liu demonstrate the value of comparative scholarship like the work collected in this book. The contemporary Chinese social and environmental situation comes into focus when principles such as "slow violence" are re-directed from areas like South Asia and West and Southern Africa, where they have been usefully applied in previous studies, to environmental justice narratives in China. As Li and Liu suggest, "Native Chinese ecocritics tend to ignore environmental justice concerns." This is not universally the case, but basically this is an accurate statement about trends in Chinese scholarship, even though there are clearly environmental justice issues in that part of the world. This new study shows graphically how contemporary China is splitting apart—the increasingly affluent, urban middle class moving in one direction, the rural poor sliding even more deeply into poverty and disease. And everywhere the natural environment, too, is a casualty of China's "success." The lens of environmental justice makes this paradox—and, some might say, "this tragedy"—painfully evident.

Perhaps corporate and government officials—and common citizens—in China would do well to heed the story of economic success and social and ecological crisis in nearby Japan. Tsutomu Takahashi's profound explanation of Michiko Ishimure's classic narrative of Minamata Disease in southern Japan, *Paradise in the Sea of Sorrow* (1969), offers a blueprint for the process by which economic development so often results in degradation of the human and ecological community—and how marginal regions (rural, distant from centralized government, inhabited by non-mainstream social groups) have time and again, throughout the world, become "sacrifice zones." While focusing on particularities of the Minamata story and Ishimure's eloquent text, Takahashi also shows in his project how readers can decode the symbolic discourse of environmental justice and postcolonial discourse that appears whenever distant cultures impose their desires and their material power, through military and/or industrial efforts, upon local communities and land or seascapes. Readers will note the acute irony with which Takahashi invokes the rubric of the Global South to describe Tokyo-centric Japan's treatment of the village of Minamata on the "backward" southern island of Kyushu. Many readers in North America and Europe might find it odd to describe any place in a country like Japan as part of the Global South, but various articles in this collection make it clear that even generally affluent nations have their neglected, marginalized, and degraded communities and ecosystems. The town of Minamata (and nearby Minamata Bay) are Japan's poster children for the Global South. The fact that Takahashi himself writes as a faculty member at Kyushi University in defense of a damaged region on his own island and in support of a classic literary work by a Kyushi-based writer is also important—a sign of "the empire writing back," to use a postcolonial catchphrase. It should be clear after reading the East Asian (Chinese and Japanese) contributions to this collection that these are not examples of Ivory Tower scholarship, mulling over timeless questions of literary and scholarly traditions. These articles are deeply—and at times aggressively—engaged in national and international politics. This work signals if not a new then at least a *growing* aspect of the environmental humanities in general—and ecocriticism in particular—the use of humanities scholarship to draw attention to sensitive social and environmental problems, to offer conceptual and hermaneutic tools for citizens and officials seeking to understand public responses to concerns and crises of all scales, and to provide at the same time Big Pictures and local examples that enable us to understand the comparative patterns across cultures and bioregions.

For instance, having contemplated several studies in which traditional postcolonial scenarios have been sketched, pitting the colonizing state against various marginalized regions and communities, we turn to Jyotirmaya Tripathy's penetrating analysis of the fundamental problem with this paradigm, specifically within the Indian context. As he says, "The postcolonial state legitimates its resolve to over-exploit environment through the narrative of development which gives it immunity not just from the international community, but also from its own people." In his study, Tripathy seeks to tease out the "multidimensional nature" of Third World environmental activism, showing "its mediation through other categories like caste and gender, its multiple and often contradictory expressions, and the unevenness in the delivery of such activism." He deflates and complicates the simple, predictable idea that "environmentalism of the poor" is merely a matter of the quest for equity. For him, the key questions include: How do competing and conflicting notions of environment and development reveal new dimensions of the human-environment relationship? How does the environmentalism in the Third World repress and shape resistance narratives? What even constitutes "environment" and "environmental problems" in the Third World? Lest readers of this volume become too comfortable with preconceived understanding of environmental activism, simply slapping First World theories and practices onto the Third World, Tripathy pulls back and retheorizes environmentalism from an Indian—and more broadly Third World—perspective.

In directing our attention to bioprospecting from an Indian perspective, Pramod Nayar accentuates the understanding that there is a fundamental difference in perspective between First and Third Worlds. While the "developed world" hungers for ever more resources in order to become ever more wealthy and comfortable, countries like India—including human and non-human beings—function as resources for the rest of the world. Nayar states: "The ecology of tissues and organs, stem cells and genetic matter within circuits of capital determines the material realities of bodies in the Global South." Opportunity and victimization collide in this representation of "biological citizenship" from a global southern point of view. The field of ecocriticism is rife with postcolonial readings and, day by day, with New Materialist interpretations of cultural texts, but Nayar's project demonstrates what happens when one engages in a "postcolonial materialist ecocritical reading"—an explosively political approach to ecocritical analysis. "We have just been corporeally globalized," announces the critic at the end of his piece. This is a striking result of the comparative approach offered in this volume through a series of regionally specific studies.

Making our way gradually westward, we turn first to South Africa as American scholar Laura White examines novelist Zakes Mda's development of "innovative ways to think about the creative potential of human movement through urban spaces." In his 1995 novel *Ways of Dying*, Mda shows the city to be what White calls a "living environment constituted through human and nonhuman interaction." While outsiders—and especially denizens of the First World—may be inclined to think of the developing world as consisting of rustic, wild spaces and other species, White argues that in expanding ecocriticism so as to accommodate the special priorities of the Global South, it is necessary to "revis[e] definitions of nature and environment to reflect that urban poverty, housing and urban development, and waste management" can be recognized as "environmental issues." As Mda reconfigures the European tradition of the ecoflaneur to suit the experiences of his black South African protagonist, who walks through the urban slums of his country, we are not treated to a rosy narrative of oncoming democracy and majority rule as a panacea, the solution to the woes of apartheid. This story, claims White, thwarts easy, predictable solutions, such as the valorization of the poor and the idea that slum-dwellers can take on themselves the problem of their own poverty. Instead, culpability for poverty and pollution is ascribed more intractably to capitalism and private property—to a deeply and globally entrenched system, not only to specific groups or communities within the country.

If ecoambiguity in the South African context includes discussion of urban poverty as an essential environmental topic, when we move northward to Ukraine we encounter instead another case of domestic colonialism—namely, the Soviet-era industrial development that resulted in one of the world's most infamous technological disasters, the 1986 nuclear accident at Chernobyl. Notably, this accident did not happen on the outskirts of Moscow, but rather in the town of Chernobyl in the republic of Ukraine, resulting in long-term ecological damage and social disruption in a community with a deep tradition of environmental consciousness. While readers throughout the world are familiar with Ukraine's nuclear reputation, few outsiders are aware that there exists both a long tradition of environmental stories and a new ecological culture in Ukraine, defying the tainted international view of this eastern European country. Ukrainian ecocritic Inna Sukhenko painstakingly details the literary tradition of her country, revealing that resistance to ecological disasters by way of artistic expression dates back well before the Chernobyl accident, at least to the 1950s. For Sukhenko, Chernobyl was a "watershed event," not only because of its impact on the land and on public health, but because of what it reveals about morality, social processes, and access to "truth." From China to Japan, India to South Africa, and also here in Ukraine, we encounter the question of how the nation-state supports, listens to, and communicates with the populace. The case of the Chernobyl accident is a particularly egregious instance of the government withholding information essential to public health. If anything good has come from Chernobyl, perhaps it is the efflorescence of literary responses—even an entire category of writing known as the "Chernobyl genre," characterized by the tension between the historicity of the nuclear accident (shown through reportorial narrative style) and the unique and varied perspectives of individual survivors (shown through various representations of subjectivity). Sukhenko argues, in her ironic conclusion, that "Ukrainians have used Chernobyl to trigger cultural self-awareness and renewed ecological attentiveness." In a sense, this is a familiar story: after disaster, we begin to wake up. Canadian singer Joni Mitchell famously crooned, "You don't know what you've got till it's gone."

Crossing the Atlantic to North America, Indian scholar Salma Monani looks at the "obliquely themed" work of Mohawk filmmaker Shelley Niro titled *Kissed by Lightning* (2009), which emphasizes human conflict rather than explicit ecological subject matter. What emerges through concentrated ecological reading of the film is a "natureculture continuum" by which a story of one woman's personal loss synecdochically comes to represent a larger cultural story of land rights. In interpreting Fourth Cinema (or Indigenous Cinema) ecocritically, Monani suggests, it is important to be sensitive to "eco-sensibility" that may not be overtly expressed. One must begin with the premise, it seems, that everything cultural is connected to nature and everything natural is connected to culture. Nature and culture are bound up in a continuum. Further, she points to a "trans-indigenous" sensibility that reaches across national boundaries and underlies indigenous experience and aesthetics in many parts of the world. Just as Tsutomu Takahashi points to the irony of identifying rural Kyushu as an underdeveloped and colonized part of the world within the territory of Japan, Monani argues that Fourth Cinema and other expressions of indigenous culture in North America deserve attention in discussions of environmental conditions in the Global South. *Kissed by Lightning* is presented here as a "narrative of alternative eco-contact" insofar as "nature has been subsumed into material goods, financial survival, and technological artifacts" that characters experience in their daily lives.

Still in the Americas, Dora Ramirez-Dhoore considers the symbolic ramifications of La Leyenda Negra (the Black Legend) that tells the story of how Mexican culture was founded through the colonial relationship between the Spanish and the local indigenous cultures. One aspect of colonial irony—if not ecoambiguity per se—that Ramirez-Dhoore points out is that fact that the Spanish have often been denigrated among European societies, but they have also strategically denigrated cultures, such as the indigenous people of Mexico, whose lands and natural resources they desired. With a particular focus on novelist Laura Esquivel's 2006 *Malinche*, Ramirez-Dhoore unpacks the geographical and religious reverberations of Spanish colonialization in the Americas, particularly showing how even contemporary Mexico bears the effects of this history. Through the novel's land and water imagery, she traces how "the Black Legend continues to redefine Mexico's environment as both usable and expendable." When "politics are connected to the Earth," she suggests, the implications are often devastating and lasting. In focusing on the lingering rhetorical effects of a legend dating back to the beginning of the 1600s, Ramirez-Dhoore shows yet another dimension of ecocritical thinking in the context of the developing world—the idea that history matters. What happened long ago continues to affect the ecology and the daily lives of ordinary citizens in once-colonized lands, even as governments and wealthier individuals seek to float free of the past and explore new opportunities for growth. The theme of rape at the center of the Black Legend uncannily fits the ongoing rape of Mexican society and the Mexican environment, no longer by Spain but by global capitalism on an international scale.

Aarti Madan's study of Brazilian author Euclides da Cunha's *The Amazon: Land Without History* (1909) maintains the focus on history, but suggests that da Cunha, who is sometimes vilified by ecocritics in Brazil for an imperialist nostalgia that acknowledges his own role in cultural and environmental destruction, was actually engaged in an effort to "transculturate mapmaking." She argues that da Cunha at once outlined prospects for clearing and populating Amazonia and, at the same time, "nurtur[ed] civilization's sylvan roots," crafting a cultural mythology of attachment to the forest that continues to shape Brazilian identity, but in a more positive way than the Black Legend functions in Mexico. The ecoambiguity unique to da Cunha's role in the development of Brazilian society is that there seems to be a direct and supportive relationship between the settlement and the preservation of the natural landscape. Madan walks a fine line in her article between identifying the unique characteristics of Brazil's environmental history and explaining Brazil's status as an illustration of how social ecology functions more broadly in the Global South.

The final article in this collection is George B. Handley's ecocritical rethinking of diasporic history and literature in the context of Latin

America and the Caribbean, with a particular eye toward correcting the limitations of traditional modes of literary regionalism. The transnational tilt of this study perfectly captures the approach of this entire collection, which aims to show the idiosyncrasies, the overlappings, and the contingent, ongoing processes of change that dictate environmental experience and expression in all corners of the world. Handley states directly that to understand Latin American literature ecocritically—and as the rest of this book demonstrates the same may well hold for literary expression from every part of the world—"it is necessary to understand the natural fact that human and natural histories are not separate genealogies but are instead ambiguously and perpectually confused." He boldly calls for "a fundamental reorientation of culture toward the natural world, one that assures that nature is protagonist not scenery." Readers who hear in this line echoes of earlier ecocritics who emphasized the ecocentric ideologies of North American nonfiction "nature writing" would not be far afield. Handley writes, somewhat against the grain of postcolonial criticism that highlights the local particularities of specific cultures that "Criticism . . . should not seek to reduce literature, like a dam to a river, to an ideologically fixed point. In comparative and ecological readings, major nature writing from the North becomes less the suspicious enemy and more a valuable point of comparative reflection on similar questions." This argument strikes right to the core of the entire collection, which juxtaposes treatments of various histories and narratives that might initially seem antithetical, such as the discussion of La Leyenda Negra's lingering damage to Mexican culture in contrast to the idea that da Cunha's 1909 essays about the Brazilian Amazon must be remembered in order to appreciate the delicate balance between settlement and preservation hinted at in that early work. It is only fitting that the final piece in this book should hit such an explicit and powerful note of cross-cultural, postnational comparison.

What do these articles all add up to? Where might readers go with these examples and these critical methodologies in mind? First, we hope to demonstrate that there are both local particularities and accessible universalities in the environmental stories of disparate societies around the world. While many students and scholars in the humanities still emphasize specific countries or regions, or perhaps specific academic disciplines, every one of the contributions to this volume demonstrates an effort to link together apparently incompatible material. The need to think expansively, even comparatively (i.e., cross culturally and interdisciplinarily), is a strong theme here and may well represent the preeminent direction in contemporary environmental humanities research and teaching. Second, scratch the surface of any culture, and you will find a fascinating, disturbing, and potentially beautiful story—sometimes these seemingly incompatible qualities go perfectly together, as in the case of

Michiko Ishimure's classic tale of small-town fishermen and their families in southern Japan, far from Tokyo, suffering from Minamata Disease because they've been practicing their traditional fishing lifestyle in a bay contaminated by chemical effluent. Whether we're talking about Zakes Mda's representation of environmental sensibility among residents of black South African slums during the country's transition from apartheid or Mexican national identity as "the Other," as the collective offspring of la Malinche "the whore" who encompasses both colonial conquest and indigenous memory, we find stories—case studies—that combine beauty and tragedy. And third, what we end up with is a picture of artists and scholars—citizens—throughout the world, all striving to engage with complex and troubling social and environmental realities by way of images, narratives, and symbols—and such engagement has political implications. All of these articles are examples of politically engaged scholarship. Through these multinational and politicized studies, we hope to promote the development, the advancement, of the community of global ecocritics.

WORKS CITED

Mitchell, Joni. "Big Yellow Taxi." *American Earth: Environmental Writing Since Thoreau.* Ed. Bill McKibben. New York: New American Library, 2008. 490. Print.

Thornber, Karen Laura. *Ecoambiguity: Environmental Crises and East Asian Literatures.* Ann Arbor: U of Michigan P, 2012. Print.

ONE

Chinese Literature and Environmental Crises

Plundering Borderlands North and South

Karen Thornber

In the *Tian shuo* (Speech of Heaven), a treatise on the relationship between Heaven and Humankind, the Chinese literatus Liu Zongyuan (773–819) transcribes a conversation with the revered poet and essayist Han Yu (768–824). After describing damage inflicted by parasites, Han Yu declares:

> People's destruction of the Primal Forces [*yuanqi*] and of Yin and Yang [yinyang] is even greater [than that of vermin]. People plow virgin plains, fell mountain forests, dig latrines, open up river channels, boundary ditches, reservoirs, and ponds . . . People make it impossible for Heaven and Earth and the myriad things to attain their true state [*deqing*] . . . In my opinion, if there were someone/something that could damage humans and make their population daily fewer and diminish by the year, and make the damage to the Primal Forces and Yin and Yang increasingly less, then such an entity would be someone/something that had achieved merit for Heaven and Earth.[1]

Han Yu here decries human destruction of nature, arguing provocatively that reducing the human population would benefit both heaven and earth. The *Speech of Heaven* depicts Liu Zongyuan as at a loss to respond to such a hypothesis.

More so than their counterparts in other areas of the world, premodern Chinese literatures, fine arts, religions, and philosophies frequently idealized abstract visions of the natural world and harmonious images of human interactions with it. But as these lines from Han Yu suggest, such representations more often defied than reflected empirical reality. Like most societies, Chinese have an extensive history of using landscapes to their own benefit. As the historian Mark Elvin has observed, for three thousand years, the Chinese cleared forests and other vegetation, terraced their nation's hills and mountains, and partitioned its valleys. They diked and dammed China's rivers, hunted its animals, and destroyed habitats in order to improve their economy. By the late imperial period, there was little left in China that could be called "natural" (321).

In the last century environmental degradation in China has only become more severe. Today, urbanization and economic growth are sending the country on a collision course with a water crisis (Yardley). More than one-fourth of the nation's land is now desert; desertification rates have doubled since the 1970s, and threats to biodiversity continue to grow (Liu Yu et al., 31). Moreover, the nation's continued reliance on coal and charcoal stoves, as well as rapidly increasing car ownership, have resulted in ever more severe air pollution. To make matters worse, approximately eighty percent of household and industrial trash in China goes untreated (Bradsher). The Chinese government has announced numerous environmental laws, regulations, and policies that when implemented have enjoyed at least limited success. But efforts to curb damage to environments are often ignored in favor of economic growth or in some cases simple survival.

For its part, Chinese literature is famous for celebrating the beauties of nature and for depicting people as intimately connected with the nonhuman world. But in fact, some premodern and much modern Chinese fiction and poetry portray people damaging if not destroying everything from small ponds to the entire planet; references to ecodegradation have appeared regularly in Chinese literature since the 1980s.

Just as with most literatures on devastated environments, one of the most striking features of Chinese literature on ecodegradation is its engagement with the phenomenon of *ecoambiguity* (environmental ambiguity). Ecoambiguity refers to the inconsistent, frequently contradictory interactions between people and the natural world and captures better these diverse relationships than the more familiar ecocritical concepts of ecocentrism, ecophilia, and ecophobia.[2] This article discusses Chinese literary negotiations of two important forms of ecoambiguity: 1) rapidly changing attitudes toward nature in Zhong Acheng's (A Cheng) novella *Shu wang* (*King of Trees*,1985), where individuals alternate between awe and indifference toward the natural world; and, related to this, 2) "loving nature to death" in Jiang Rong's (Lü Jiamin) novel *Lang tuteng* (*Wolf Totem*, 2004), where the same person/people who show obvious fascination with nature, or who go so far as to declare that they respect or even love nature, contribute, deliberately or inadvertently, to damaging or destroying the nonhuman. Although *King of Trees* and *Wolf Totem* do not focus on China's environmental exploitation of lands outside its borders, including the Global South, both works highlight the damage Chinese inflicted along their own frontiers during the Cultural Revolution (1966-76), a haunting precursor of contemporary policies of southward expansion.[3]

King of Trees is based on A Cheng's own experiences during the Cultural Revolution in the Xishuangbanna, a Daizu Autonomous Region in the Chinese deep south bordering Laos and Burma and one of China's most diverse biological sanctuaries; A Cheng was one of approximately 200,000 educated youth (*zhiqing*) sent to the Xishuangbanna, where more than 600,000 acres were rapidly felled, changing even the region's climate.[4] The text reveals the conflicting attitudes toward trees and landscapes more generally held by young men under orders to mutilate unfamiliar, yet awe-inspiring ecosystems that authorities have condemned as "old" and "useless." A Cheng's novella demonstrates how readily, and without explanation, awe can become indifference, only to be transformed into terror at what has been done in the name of indifference.

Narrated by a young man who does not recognize this contradiction within either himself or others, *King of Trees* underscores the difficulties of acknowledging, much less addressing, enablers of large-scale environmental destruction. The novella touches on numerous differences in attitude vis-à-vis the nonhuman. The clearest contradictions are those between the characters Li Li and Xiao Geda (Knotty Xiao). Li Li is an enthusiastic revolutionary who has been dispatched to the mountains, while Knotty Xiao is part of a production team that has been working there for nearly a decade. The principal task of the newly arrived workers is to help replace trees condemned as useless with "useful trees" (*youyong de shu*).[5] Li Li believes that existing vegetation, as an "old thing" like the actual four olds (*sijiu*; old customs, old habits, old culture, and old thinking, the destruction of which was one of the stated goals of the Cultural Revolution), must be removed at all costs. His behaviors are in line with his feelings, and he quickly proves himself the group's most dedicated logger. In contrast, Knotty Xiao, already uncomfortable with the continuing destruction of trees and undergrowth on the mountains, is deeply disturbed by Li Li's determination to cut down even Shu wang (King of Trees), the region's largest and most impressive tree, one that Knotty Xiao and many locals believe sacred; the death of this tree at the hands of Li Li and his friends is followed quickly by Knotty Xiao's own, so pained is he by its demise.

The attitudes of the first-person narrator and most of the educated youth outside Li Li and his inner circle fall between those of Li Li and Knotty Xiao. These young men appreciate, indeed stand in wonder of, the rural landscape. On the other hand, they have few qualms about participating in its destruction. Despite their initial fascination with this mountainous region, it takes felling the King of Trees and transforming mountains into raging infernos for them to become uncomfortable with human alteration of ecosystems. Although it is not surprising that they obey the order to clear the land, since they will be severely punished if they do not, nothing prevents them from voicing internal outrage about the loss of so much vegetation. The narrator does not address the contradictions in his or the other young men's attitudes, nor does he even appear to be aware of them. The *King of Trees* demonstrates how easily appreciation and reverence of ecosystems can be accompanied by indifference to them and even satisfaction at their destruction. The immediacy of reverence when encountering a space is contrasted with eagerness to trash it; second thoughts about this destruction arrive only belatedly.

King of Trees opens with a group of educated youth arriving in a valley, awestruck by the wild landscape by which they are surrounded; when they learn that this is their final destination, they leap to the ground in excitement, one after the other (72). The narrator's language stresses both the ruggedness of the land and the joy of the young men's encounter with it. The following day, the educated youth climb one of the mountains with their new team leader. Reaching the top, they look out at the scenery and are rendered speechless: "All that remained of the distant mountains was their color, undulating blue spreading out, growing paler with each tier. We all gasped dumbly, opening our mouths one after another to speak but at a loss for words. I suddenly felt as though these mountains were like the ripples of the human brain. I just didn't know what they were thinking" (82). The final three sentences of this quotation powerfully juxtapose rows of actively thinking mountains with rows of young men who are "gasping dumbly" (*daidai de chuanqi*) and "cannot speak" (*shuo bu chu hua*) despite their mouths having opened one by one to do so. Just as in the opening paragraph they jumped off the tractor one by one, now they open their mouths in succession. Yet having become even more mesmerized by the mountains, the youths are rendered speechless, writing capturing what verbal discourse cannot. Even more important, the mountains are presumed to be actively thinking while the young men's brains appear frozen.

Yet fascination with their surroundings, even curiosity about the consciousness of their surroundings, does not stop the young men from being swept along by rhetoric clamoring to reshape it. As everyone is "gasping dumbly" and the narrator ponders what the anthropomorphized mountains might be thinking, the team leader announces that they have been tasked with helping clear more than 1,600 acres of mountain land and replanting this space with "useful trees." After Li Li determines that all the mountains within view will be planted with "useful trees," he responds that this is magnificent, that transforming China is just magnificent (82), a sentiment with which the other young men concur. And they do not appear at all troubled when the team leader provides additional details, when he tells them that they are to topple the trees growing on the mountain where they are currently standing, and that they will burn its timber, terrace it, excavate additional holes, and ultimately plant useful trees (82). In fact, rather than questioning whether such extensive manipulation of the mountain's ecosystems is really necessary, the young men immediately ask why the large tree dominating a nearly replanted neighboring mountain has not been felled. When the team leader responds that it has "become a spirit" and that real trouble awaits whoever cuts it down, the young men laugh and ask how this is possible.

Similar dynamics are repeated and intensified throughout *King of Trees*. With the notable exception of Li Li and a few of his close friends, the young men remain in awe of their surroundings. Just hours after agreeing that destroying mountain ecosystems, including large trees, is a magnificent task, their curiosity gets the better of them, and they decide to climb another mountain to take a look at a tree they had seen from a distance earlier that morning. Discovering that this is the King of Trees, the young men again find themselves rendered speechless, their minds empty. The narrator comments that he could neither speak nor sing and that everyone looked at each other, swallowing hard and backing up (85). Although the narrator's mind is blank for only a short while, it is a long time before the young men are able to release themselves from the spell of the tree. The tree, on the other hand, retains its voice; the narrator describes its leaves as gurgling, making sounds as though it were talking to itself or playing with large groups of children (87). As before, the nonhuman takes on human voice while people stand wordless.

Interestingly, however, when the time comes to begin felling trees—just a day after the incident described above—the landscape appears to have lost its magic. The space that fewer than twenty-four hours before had left the narrator at a loss for words now is described relatively objectively, the narrator noting that their job was naturally [*ziran*] to cut trees. Because, allegedly, no one had touched the forest for hundreds of thousands of years, the entire woodland had grown into a single mass, the trees competing with one another for sunlight and vines creeping from tree to tree, every year thickening the crust of vegetation (88). In the pages that follow the narrator calmly outlines the difficulties of felling trees. He notes that the forest has never been touched by human hands, yet he does so not out of anger at the loss of primeval species but instead to impress on the reader the physical labor required to eradicate them.

Notably, the only thing that seems to bother the young men about destroying the forest is the tedium and the arduousness of the task. They eagerly join Li Li in his efforts to cut down a colossal tree not unlike the King of Trees even while knowing that previous loggers have avoided this flora because the time required to cut it down would not justify the credit they would receive for so doing. To be sure, by concentrating their efforts on this one tree they are sparing a considerable number of smaller ones. But more significant is the satisfaction, indeed pleasure the young men derive from felling something so massive. The moments following the tree's fall find them somewhat shaken, but these feelings quickly dissipate, and the young men walk jovially down the mountain. The narrator contrasts the demeanour of the young men with that of Knotty Xiao, who having offered them eleventh-hour assistance now is visibly distressed.

Emotions change with the felling of the King of Trees. Although he had been very eager to participate in destroying the region's other massive tree, the narrator now reveals himself as deeply torn, something likely true of many others in the valley and up in the mountains; he claims that his heart and mind were in turmoil, and that he can't determine whether cutting the tree is right or wrong (120). It is only now, with conditions bordering on the extreme, that he stops to ponder the morality of the group's behaviors.

Agitation increases after the forest is set ablaze to clear it for farmland. As the flames spread fiercely up the mountain, the narrator, other educated youth, and villagers alike begin to empathize with the trees, the narrator commenting simply, "The sight of the trees all silently lying there made people feel anxious for them" (122). Resembling a boiling cauldron, the mountain in *King of Trees* screams out in pain; its surface scalded, it "released all kinds of strange cries, an entire universe alarmed" (124). For a moment Li Li and his friends remain indifferent toward the destruction surrounding them; they talk and laugh together as though nothing has happened. But soon they too are affected. As the narrator observes, "The conflagration burned everyone's spirits" (126). But by then it is too late.

Creative texts that discuss people who damage environments regularly portray them as demonstrating conflicting attitudes toward the nonhuman. Most obvious are the differences in beliefs, emotions, and perceptions between seemingly cohesive groups. More complex but no less frequent is the environmental ambivalence of specific individuals and communities. A Cheng's novella *King of Trees* warns how rapidly feelings about the environment can change and also about the sometimes deadly consequences.

A second and related form of environmental ambiguity articulated in Chinese and other creative writing is the phenomenon of loving nature to death. Environmental degradation would be difficult enough to contain if the attitudes propelling it were entirely and obviously ecophobic. But damage to landscapes is even harder to prevent and remediate because people's basic sympathy toward the nonhuman regularly accompanies behaviors that inflict harm (whether or not expected), or at least do little to prevent harm. Popular environmental discourse often underscores the need to develop feelings of respect for other species, draw closer to the nonhuman, and even learn to "love nature." Sometimes these emotions translate directly into actions that benefit the planet's ecosystems. Yet many literary works argue that little prevents well-meaning individuals and even self-proclaimed environmentalists from acting in ways that can ultimately damage the nonhuman. These types of conflicts between attitudes and behaviors are prevalent throughout twentieth- and twenty-first-century East Asian literatures, including in Jiang Rong's overnight sensation *Wolf Totem.*

This novel is based on its author's experiences in Inner Mongolia between 1967 and 1978, the Cultural Revolution and its immediate aftermath. It follows Chen Zhen, a young Beijing intellectual who during the Cultural Revolution is sent to Inner Mongolia's Olonbulag as part of a production team and becomes enthralled with the region's human and nonhuman occupants.[6] While in Inner Mongolia, Chen Zhen witnesses an influx of Han Chinese (China's most populous ethnic group) into the grasslands there, the resulting transformation of the landscape from a space of nomadic herding to grain cropping, and the extermination of the region's wolves. Decades later, when he and his friend Yang Ke return to Inner Mongolia, they discover that much of the grassland has become desert. Particles from this arid region in fact have spread well beyond the Olonbulag; *Wolf Totem* concludes in 2002 with Beijing shrouded in sand.

Jiang Rong's novel is replete with environmental ambiguity, particularly the disjunctures between human attitudes and behaviors. Most notably, Chen Zhen imprisons and ultimately kills a wolf to which he has become deeply attached. Unlike the vast majority of Han Chinese, Chen Zhen is fascinated by wolves. Averting a wolf attack not long after his arrival on the grasslands, and fascinated by the orderly retreat of the animals, he is described as falling under their spell, feeling within a combination of fear, reverence, awe, infatuation, and obsession with wolves. For Chen Zhen, the narrator claims, the Mongolian wolf was not a living thing that merely had touched his soul; instead, it was a living thing that had struck his soul (9). His soul not just touched (*chuji*), but struck (*jichu*), Chen Zhen is completely mesmerized. One night, after watching a dramatic struggle between a wolf on the one hand and domestic animals and their human owners on the other, he decides to learn as much as he can about these impressive creatures and imagines raising one himself. He finds the perfect opportunity during his second year in Inner Mongolia when he and his friends raid a wolf den and bring home seven pups. Finally holding a wolf, Chen Zhen declares the animal to be the most noble and valuable, indeed the most precious small life on the Mongolian grassland (160). He decides to keep the strongest pup for himself and raise it to adulthood.

Chen Zhen is aware of some of the many contradictions between his feelings toward the wolves and his treatment of them. He truly admires the animals and hopes that increased contact with wolves will lead to greater understanding of them. But *Wolf Totem* soon reveals the significant disjuncture between Chen Zhen's reverence for the animals and his mistreatment of them.

After arriving home, the young men determine that their Mongol friend will kill the unclaimed cubs. Chen Zhen merely sighs deeply, "There's no other way" (168). Watching the murder of the first wolf is almost more than he can tolerate. But still he does nothing. The narrative makes it clear that despite his protests to the contrary, his attraction to wolves in the abstract, combined with little thought about the well-being of individual wolves and then with simple cowardice, leads directly to the demise of five pups.

The contradictions continue as Chen Zhen begins to raise his wolf. He believes that nurturing the pup will help him and ultimately Han Chinese gain a deeper understanding of wolves; this will be no small accomplishment, since the Han Chinese allegedly have feared and tormented these animals for thousands of years. Yet his preoccupation with a distant future of understanding not only wolves, or even simply wolves and people, but also both animals and relationships between animals and people blinds him to established knowledge and present conditions, which if not taken into consideration make the future he envisions nearly impossible.

Enthralled by his quickly growing pup, Chen Zhen refuses to listen to his Mongolian friends. And although initially enthusiastic about helping Chen Zhen raise the wolf, his friend Yang Ke soon recognizes that this is not a viable enterprise and expresses dismay at the inhumanity of Chen Zhen's treatment of the pup. Yang Ke declares that it causes him too much grief to see a little wolf tied up all day like a little prisoner. This is because, he claims, the wolf loves freedom more than any other animal. Hearing these words, Chen Zhen is conflicted. He of course thinks of returning the wolf to the wild, but he believes that he cannot now release the wolf, there being too much about this animal that he does not yet understand (350). Likening the chained wolf to a prisoner and disturbed by what he sees, Yang Ke declares pitiful the spectacle of this incarcerated animal. More significant, he observes that wolves love freedom more than any other animal, making Chen Zhen's treatment of this creature especially "blasphemous" (*xiedu*).

Chen Zhen's rhetoric is grand and his aspirations lofty, yet his behaviour is anything but. There is no evidence that keeping the pup in shackles will allow him to understand its "true nature," much less the "true nature" of wolves. On the contrary, there is considerable likelihood of the opposite: keeping a wolf in captivity, particularly under such deplorable conditions, probably will teach him far less about wolves in their "natural" state than about how they react to internment. This, paradoxically, is the opposite of Chen Zhen's aspirations.

Not surprisingly, the larger the wolf grows, the more frustrated it becomes with its chain, pulling and yanking it so desperately that it seems as though the wolf will not stop, that it ultimately will strangle itself to death (541). Even though the wolf continues to fight its captivity and its injuries become increasingly severe, even though Chen Zhen admits he knows just how cruel it is to imprison this animal, he refuses to release it into the wild, claiming that to do so would be lethal. Rather than give the animal even a moment of freedom, Chen Zhen keeps it imprisoned to the bitter end, until finally, after enjoying one last moment together, he has no choice but to kill the animal he had hoped to "raise to old age." Chen Zhen falls to the ground in total despair, tears streaming down his face; he embraces the pup, and touches its forehead and neck against his own, one final time. Then he grasps a spade, and holding it behind him runs back to the wolf, and smashes it on the back of its head.

> The wolf didn't make a sound, softly falling to the ground, like a true Mongolian grassland wolf, holding out until the very end—At that moment Chen Zhen felt as though his soul had been knocked out of him. He seemed to hear again the clanking sound of his soul rushing out of the top of his head. This time it seemed as though the soul that flew out would never again return. (548)

The narrator accentuates the increasingly close bond between man and wolf—the two first enjoy friendly physical contact (licking hands, rubbing knees, scratching and pouncing on stomachs), then they mix bodily fluids (blood and tears), and finally they entangle souls, as both are destroyed.

But, even more important, *Wolf Totem* shows Chen Zhen's intoxication with wolves and insistence on understanding these animals coming at the expense of their lives; the man who had described his soul as being struck by this animal himself clubs one to death. There is no question of his great respect for wolves as an abstract concept and even as an actual species, but relationships with individual wolves prove far more complicated. Fascination with animals not only coexists with but also frequently results in harming them. His acts have consequences just as deadly, albeit on a much smaller scale, as the Han Chinese invasion of Inner Mongolia and are suggestions of those as China expands deeper into the Global South. Far from resolving the ecological conundrums it describes, most literature on environmental distress further ambiguates these ambiguities, bringing to light deeper discrepancies. In fact, poetry and prose regularly counter the unstable, inverted pyramids of environmental Ponzi schemes with their own mushrooming miasmas of environmental ambiguity. Creative writing often makes matters even more confusing by remaining silent about its uncertainties: many narrators and characters seem unaware of the discrepancies they depict or exhibit. Indeed, to the extent that they are aware of them, people often suppress the contradictions of their interactions with environments. Analyses of literature can provide sharp lenses for deeper insight into processes of environmental degradation by highlighting creative articulations of these disjunctions. Better understanding of the uncertainties and contingencies of ecodegradation in literature, and in life, should allow for more productive relationships between people and the natural world.

WORKS CITED

A Cheng. *Shu wang. Qi wang, shu wang, haizi wang*. Taipei: Dadi Chubanshe, 2007. 71–130. Print.

Bosshard, Peter. "China's Overseas Dam Builders: From Rogue Players to Responsible Actors?" *Asia-Pacific Journal*. Web. 26 April 2010.

Bradsher, Keith. "Earth-Friendly Elements, Mined Destructively." *New York Times* 3 July 2009: B1, B5. Print.

"China Boss in Peru on $50 Billion Peak Bought for $810 Million," *Bloomberg Businessweek*. 1 November 2010. Web. Accessed July 22, 2013.

Diamond, Jared. *Collapse: How Societies Choose to Fail or Succeed*. New York: Viking, 2005. Print.

Elvin, Mark. *The Retreat of the Elephants: An Environmental History of China*. New Haven: Yale UP, 2004. Print.

French, Howard W. "The Next Empire?" *Atlantic* 305:4 (2010). 59–69. Print.

Jiang Rong. *Lang tuteng*. Taipei: Fengyun Shidai, 2005. Print.

Lamont, H. G. "An Early Ninth Century Debate on Heaven: Liu Tsung-yüan's *T'ien shuo* and Liu Yü-his's *T'ien lun*." *Asia Major* 19:1 (1974). 37–85. Print.

Liu Yu et al. "The Politics and Ethics of Going Green in China: Air Pollution Control in Benxi City and Wetland Preservation in the Sanjiang Plain." *Forging Environmentalism*. Ed. Joanne Bauer. Armonk, NY: M. E. Sharpe, 2006. 31–101. Print.

Shapiro, Judith. *Mao's War against Nature: Politics and the Environment in Revolutionary China*. New York: Cambridge UP, 2001. Print.

Taylor, Ian. *China's New Role in Africa*. Boulder: Lynne Rienner, 2009. Print.

Thornber, Karen Laura. *Ecoambiguity: East Asian Literatures and Environmental Crises*. Ann Arbor: U of Michigan P, 2012. Print.

Yardley, Jim. "Beneath Booming Cities, China's Future is Drying Up." *New York Times* online. 28 September 2007. Web. Accessed July 22, 2013.

NOTES

1. Thank you to Steve Owen for the reference and the translation, adapted from Lamont, 67.

2. See Thornber (1–31) for an extensive discussion of ecoambiguity, particularly as manifested in East Asian and world literatures.

3. From Afghanistan to Africa and South America, Chinese are tearing down mountains and relocating entire towns to mine copper and other minerals ("China Boss in Peru"). Chinese have been farming in Africa for years, and in the mid-2000s the Chinese government began prioritizing large-scale agricultural investment in the continent with the hopes of increasing Chinese food security (French). Even more significant is China's expansive investment in African oil extraction to accommodate the nation's ever-increasing energy needs (Taylor 42–62). Chinese are involved in dam projects in nearly fifty countries, from Myanmar (Burma) to the Sudan, Gabon, and Ecuador (Bosshard). China is the world's greatest contributor of ozone-depleting substances and one of its major emitters of carbon dioxide; it also is one of the leading importers of timber from tropical rain forests (Diamond 359).

4. See Shapiro (159–93) for more on environmental damage during the Cultural Revolution.

5. *King of Trees* here inverts Zhuangzi's parable of the tree, where "useless" trees are spared precisely because of their uselessness.

6. The Ölonbulag is directly north of Beijing; it shares a border with Outer Mongolia and historically was the southern passage between Manchuria and Mongolia.

TWO

Tibet, a *Topos* in Ecopolitics of the Global South

Gang Yue

A. INTRODUCTION

Once upon a time the geographical location of Tibet was a blessing to the ecosystem on the Plateau as its towering mountains and forbidding terrains stood as a curse to intruders. The Himalayas served as a kind of natural Great Wall against "gate-crashers" since the British conquest of India. Not until after Sarat Chandra Das, a British-Indian pundit, completed his second journey in 1882 did the world have a functional map of Central Tibet—about one half of the Plateau under the direct rule of Lhasa. The peculiar relations of knowledge and power or exploration and conquest in the colonial era were best captured in Das's double role as a scholar and a spy. Isolation thus gave rise to a dialect of "archive and utopia"(see Richards), the cartographical endeavors for knowledge of the unknown as long as it remains unmapped. The epistemological conquest of the unknown further spawned a romantic fantasy of the mysterious, even in the 1904 British military invasion of Tibet. The "thrill of adventure," "the romance attach[ed] to the unknown," and the "curiosity to trample out the secrets of the Grand Lama's palace" was an unannounced mission for the Younghusband expedition into Tibet, beyond what London had authorized (Schell 191). It is around this utopian desire that the Shangri-La myth was developed and acted out in a lasting imaginary of Tibet.

On the northern and eastern frontiers of the Tibetan Plateau, the terrains may not have been as forbidding as the Himalayas were on the

South and West. The greater accessibility even allowed the rising Tibetan Empire to conquer the capital city of Tang China in the seventh century, long before the Mongol horsemen brought Central Tibet into the imperial fold about six hundred years later. For most part of the history afterwards, Tibet's political submission to the Chinese empire remained nominal largely because its geography would have inflicted too heavy a toll on the imperial center had the latter attempted direct rule. This geopolitical characteristic was recognized even by the powerful and aggressive rulers of the new Communist China. Instead of attacking Lhasa and managing the affairs of Central Tibet that would have created a logistic nightmare for the Chinese military, Beijing sought to compel the Tibetan delegates to sign an agreement for the "peaceful liberation" of Tibet—a legal arrangement normally reserved for sovereign nations. During the first three decades of Chinese rule, the Plateau for the most part was spared the effects of extractive industrialization and population inflows, with the exception of deforestation on the eastern edges dating back to the 1960s. The central government was, as Kenneth Pomeranz puts it, "more ambivalent about rapid development in the far west than it is today; leaders tended towards more paternalistic policies, avoiding radical cultural change, as offering the best formula for political stability there" (17).

Economic development since the 1980s has radically altered the ecological and cultural landscapes of the Plateau. The previous curse of geography to external invasion has given way to a new curse of "uneven development." This essay attempts to sketch out a contour of the current ecological crisis on the Plateau. As used in my title, "Tibet" here refers to a *topos*, a place *and* a topic that must be re-centered on indigenous voices across the global environmental scene. While Tibet remains part of the Global South both inside and outside of the People's Republic of China (PRC), it also transcends the north-south divide to stand as the Third Pole of the world. Once an adventurers' domain of fantasy and conquest, the Third Pole has recently been realigned with an environmentalist discourse and ecopolitics. Symbolically, the Third Pole is at once a victim of extractive development and an agent for environmental protection and cultural preservation. Indeed, Tibet should be conceived as a new source of inspiration for an ecocriticism of the Global South. This new locus lies in Kekexili, beyond the Western mythology of Shangri-La.

This essay sets out with the politics of "silence" surrounding a recent landslide of massive death toll at a copper mine in Central Tibet. Surrounding this environmental disaster exists a paradox of political repression and ideological appropriation of environmental universalism. It simultaneously conceals and reveals an even larger truth as we dig deeper, in the second section, into the global political economy through the dubious financing of the mining company. The third section moves our discussion to a national scene of environmental protest in the PRC during the recent decades and shows that uneven development is coupled with an equally uneven environmental politics in favor of the vocal minority of urbanites, the "first world" of the PRC. This detour of the national scene of environmental protest is traced yet to its source of inspiration from Kekexili, a no-man's land in the Tibetan heartland and the site of a fierce battle to protect the last Tibetan antelopes two decades ago. In this last section, Kekexili supersedes Shangri-La as the new synecdoche for the Plateau. It stands for a series of "firsts" in the environmental movement of China, represents an emerging Tibetan agency of neither traditional Buddhist pacifism nor modern Deep Ecology, and yet reflects none other than a Tibetan consciousness of resistance. Furthermore, the global circulation of shatoosh, the soft wool that comes from the mass skinning of the protected Tibetan antelope on the Plateau, to the world markets of high fashion maps out a new eco-political cartography of the divided yet connected world. In the epilogue I attempt to highlight the historical significance and political meaning of native resistance on the environmental front, arguing for an ecocriticism in cultural analysis that must be grounded in history, account for the broad changes of political economy and globalization, and consistent with the values of bio-cultural diversity in content and form.

Since most of the major events discussed in this essay have rarely been studied in academia, at least not in the English-speaking world, I will embed my analysis in a descriptive narrative in the hope that it will be more accessible to the reader. To keep the balance, I at times must sacrifice theoretical sophistication in favor of contextual elucidation. As a result this essay should be considered as an attempt aimed at contributing to the "infrastructural development" for an ecocriticism of the Global South.

B. THE SILENCE AROUND TIBET'S ECOLOGICAL CRISIS

On March 29, 2013, a major landslide buried eighty-three miners at Gyama Copper Polymetallic Mine in Gyama valley (Ch.: Jiama gou), about sixty-eight kilometres to the east of the Tibetan capital Lhasa. Among the dead two were Tibetan; the rest came from nearby provinces of China. They were all laborers who were sleeping in their tents when the disaster hit around dawn. The location is one of the major historic sites in Tibetan culture. Most significantly, it is the birth place of Songtsen Gampo (circa 605-650 AD), the Dharma king who founded the ancient Tibetan Empire. Home to many sacred historical temples dating back to as early as 1181, the Gyama valley attracts thousands of Tibetans every year for pilgrimage.[1]

The area was immediately sealed off to outsiders following the 2013 landslide. China's media outlets were instructed to use only the wire service of the official Xinhua News Agency, a common practice adopted in the last decade for damage control and "stability maintenance." Some netizens made brief but unsuccessful attempts to link the disaster to a broad ecological crisis and social displacement. On her well-followed blog, *Invisible Tibet,* Tsering Woeser, a Beijing-based Tibetan author and an influential dissident, began to follow the mining development in 2007 and has since exposed its negative ecological, social, and cultural impacts. Woeser was in a position to make a prompt response just one day after the landslide on account of her extensive research in the area. She forcefully disputes the official position that the landslide is a "natural disaster." She also admits that she was "dumbfounded" that there should be a mining "accident" near Lhasa, implying that the poor record of mining safety used to be concentrated in the rest of the PRC but has finally invaded the Tibetan heartland. What is at stake for her is the official attempt to cover up an environmental disaster to hide the destruction of the ecosystem and culture caused by mining on the Plateau (see Weise).

It is in this broad political context that "the silence about Tibet's ecological crisis"—the title of a recent article published on *The Atlantic* website—may well serve as an apt metaphor for an ideology of environmental universalism and its critique (Huang). The article is disturbing and illuminating at the same time since universalist ideology simultaneously conceals power relations of the global political economy and also reveals its historical and rhetorical contradictions. "Tibet is still a very sensitive topic, even if your story is about the environment and not politics," an anonymous editor of the liberal *Southern Weekly* is quoted as saying (qtd. in Huang). Five years after the protest movement that spread over the Plateau, and with more than one hundred self-immolations since (and counting), Tibet "still" exists as a *silenced topos* off limits to any meaningful discussion and debate inside the PRC as long as the issue can be (mis)construed, even remotely, as political. *The Atlantic* article, especially the views expressed by its Han Chinese author and informants, is also illustrative of their attempt to reduce a political issue to a "mere" problem of "[h]ow mistrust and fear between Beijing and Tibetans are making a bad environmental problem worse"—the subtitle of the article. As the author puts it, "China faces serious environmental challenges nationwide, and Tibet is no exception" (Huang) That statement is fair enough from a technical standpoint. What makes environmental challenges in Tibet an exception or taboo is that the *topos* has been overly politicized. After enumerating a series of mining, hydropower damming, and other problems and citing a good number of Tibetan and Western sources, the article shifts to a Chinese point of view—and I cite it at some length in the following:

> According to Wu, an environmental journalist inside China who wished to remain anonymous, the environmental problems faced by

> Tibet are similar to problems in other parts of China, and *more universal than political.* "The Han Chinese did not destroy Tibet's environment for political reasons," remarked Wu. "They are already diligently destroying their own environment." (Huang; emphasis added)

The point made here is that resolving environmental problems is a universal endeavor that transcends national and regional divides while politics remains local. The Chinese can be absolved from what they do to the ecosystem, the embedded social structures and cultural heritage in Tibet simply because they have already accomplished the same thing in their own homeland. To take on the environmental challenges in Tibet one must put it back on a national agenda for action. The article continues:

> "Different parties, including both the Chinese government and overseas 'human rights' activists, always politicize problems in Tibet, making real environmental challenges untouchable," said Gao, an environmental NGO worker in western China. He was extremely reluctant to disclose his information, worrying that any news stories coming out would pose further challenges to his NGO's efforts to enhance environmental protection. (Huang)

Caught in between the powerful state apparatus of China and the toothless and sometimes divided overseas rights groups, the self-appointed frontline guardians of Tibet's natural environment and culture should be given due respect and sympathy. Albeit my critique of their ideological blind spots, Chinese environmentalists are an indispensable force on the Plateau precisely because of their pragmatic, "apolitical" stance. Since the Communist Party is likely to stay in power for the foreseeable future, it would be wise to take a more practical approach and regain some foothold on the ground.

Indeed, that seems to be the editorial message of *The Atlantic*. Above the article is a picture of the spectacular Potala Palace, with the vast paved square in front that only old timers of Lhasa may find estranging. Two quotes are taken out of the text to highlight the point. The first one reads: "Tibet's traditional culture is more threatened by global commercialization than it is by the Han Chinese" (qtd. in Huang). This statement is consistent with the theme of environmental universalism that the Chinese is doing the same damage to Tibet as they has done to their own homeland, not for political reasons as overseas rights groups claim. Following the same logic the editorial piece takes one step further to identify the ultimate villain, namely, "global commercialization." In this way the universal impact of "global commercialization" upon the environment calls for a pan-PRC solidarity of environmental universalism that transcends the alleged political divide between Tibetans and the Han Chinese. The second highlighted quote is from the afore-cited Mr. Wu, the journalist: "Many Han Chinese in Tibet think that they understand

Tibetans. But the truth is that Tibetans have learned what to say to Han Chinese." At least Mr. Wu and some members of the new generation of Chinese environmentalists have done their homework to acknowledge the unequal political situations that exist between the Tibetans and the Han Chinese and recognize the most basic truth of human communication that must be based on political equality.

The point that I want to make here is that, while humans share the planet earth and the responsibility to protect it, the impact of the ecological crisis and the burden of protection are not equally shared due to the vast differences in power relations, uneven development, and resource distribution between communities, regions, and nations. On a more practical note, the political repression on the Plateau has made the *topos* of Tibet a more "political" or politicized place and topic; yet at the same time that a tactical insistence upon a universal environmentalism may indeed help reduce the political tension by depoliticizing the local situation. Manifest in this Chinese-Tibetan entanglement is a paradox of political repression and an ideological appropriation of environmental universalism.

The silence, it turns out, is that Tibetans inside the PRC are fearful of telling the truth, and most Chinese are not hearing their grievances or are not making an effort to listen. The buried laborers and the scarred mountains in the Gyama valley now appear to embody the silence of the repressed political reality *and* political unconsciousness of Woeser's "invisible Tibet."

C. A SIMPLE DIGGING INTO THE CIRCUITS OF GLOBAL CAPITAL

Jiama [Tibetan: Gyama] Copper Polymetallic Mine is one of the two mining operations owned by China Gold International Resources Ltd. (CGIR). The company is listed on Toronto and Hong Kong stock exchanges and has its headquarters in Vancouver, a smart choice of strategic location for the advanced mining technology and financing in Canada. The other CGIR-controlled mining operation is also located in the periphery of the historical Chinese empire, in Inner Mongolia. According to the company's 2013 First Quarter report, all revenue from gold extraction and processing comes from its Inner Mongolian operation and counts for 62 percent of the total. The Gyama operation makes up the remaining 38 percent, which entirely comes from copper mining.[2] This concentration on copper (with other derivative minerals) may explain why the Gyama mine survived the ban on gold extraction imposed by the government of the Tibet Autonomous Region (TAR) in 2007. And the name of CGIR's parent company, China National Gold Group (CNG), may have led some to mistake it for a gold mine.

On CGIR's official website, the company claims that it is "the flagship and the only overseas listing vehicle of the largest gold producer in China and 39 percent shareholder, China National Gold (CNG)" ("Corporate Overview"). To highlight its institutional advantage in connection to its parent company, a giant state monopoly, CGIR emphasizes its leveraging of "the technology, engineering, operating, government relationships, financing capabilities and international project pipeline of **China National Gold**" (original bold). Of the nine largest shareholders next to CNG on the Top 10 Shareholders list (updated on May 22, 2013), six are based in North America, one each in Europe, Hong Kong, and mainland China respectively. These include private equity funds, mutual funds, and a public pension fund from Canada. The company does not list smaller shareholders, but it is safe to imagine that they are mostly global investors from outside of the PRC, judging from the Top 10 list. The company is extremely profitable; "*Business in Vancouver* named China Gold International as **the fastest growing enterprise in British Columbia, Canada**, by percentage growth in revenue between 2007 and 2011" (qtd. in "Corporate Overview"; original bold). The rate of growth clearly benefited from the low labor, social, and environmental costs of the two mining operations in the PRC as well as the increase in price of precious metals and other commodities in the past decade. And given the 39 percent control by China National Gold, it is likely that more than half of the profits go to international investors ("Corporate Overview"). It has done well in other areas, too. In 2011, the PRC Ministry of Land and Resources selected CGIR as a "pilot of the National Green Mines Project" and also a "Model enterprise of national unity and progress" (see Environment & Development Desk 4).

Digging just a bit deeper into the history, however, the story becomes more nebulous and dark. On the official website of Tibet Huatailong Mining Development Co. Ltd (homepage updated May 23, 2013)—the company that actually runs the Gyama valley operation, the corporate structure (dated December 28, 2009) had altogether five layers of shareholder control. Tibet Huatailong was a joint venture between China Gold Group (Hong Kong) and Rapid Result Investment (RRI, based in the British Virgin Islands). The joint ownership is virtually split between 51 percent by the former and 49 percent by the latter. Since China National Gold is the parent holding company of China Gold Group (Hong Kong), it controls 51 percent of Tibet Huatailong (to conform to China's policy on domestic majority ownership in the mining industry). This simple chart of a cumbersome corporate structure is available only in Chinese and apparently outdated. But it does tell something about the financing history of the Gyama mine.

According to the report published by Environment & Development Desk (EDD), Department of Information and International Relations, Central Tibetan Administration (henceforth EDD report), there were four separate mining license areas operating in the Gyama valley and surrounding areas before April 2007. On an agreement between the TAR government and China National Gold, these separate entities were consolidated into Gyama Copper Polymetallic Mine. This is consistent with the national policy on consolidation and restructuring of the mining industry aimed to develop an economy of scale and to better implement government regulations on labor safety, social responsibility, and environmental protection. In 2010, CGIR purchased its current 51 percent of the assets from the private holding company RRI, at a total price tag about USD 590,000,000. Based on leaked emails and journalistic investigations, the EDD report asserts that "those setting up offshore entities were mostly from Chinese mainland, HK and Taiwan." It further suggests that "it could be possible that RRI [is] owned by various individuals and family trust most likely connected with the CCP in TAR and Beijing, who amassed huge wealth through the transfer of ownership of Gyama mine (13)." Irrespective of the question regarding who controls RRI, its off-shore investors have made an enormous fortune, probably fleecing the Gyama valley mine twice—first during the neoliberal period of privatization before 2007; second by selling 51 percent of its assets to the state-owned CNG for an astronomical sum.

Details of such dubious financing, deal-making, and rent-seeking will most likely remain concealed in secretive off-shore accounts. The present shareholders of CGIR—China National Gold and the international shareholders—would likely be happy to ignore the past, not to allow the hidden ghost of possible fraud to surface and derail the high-speed growth, and keep the books of the restructured CGIR clean for Canadian and Hong Kong regulators. What has been buried in Gyama valley thus has to be sealed for good, twice over.

Eighty-three deaths were bad enough for CGIR and the TAR government. Since late 1990s Beijing has persistently cracked down on illegal mining and irregular practices in the industry, reducing fatalities of more than seven thousand per year a decade ago to thirteen hundred in 2012 (see China Labor Bulletin). It should be noted that CGIR paid one million RMB (about USD 163,000) to each family of the dead (*China Merchant Daily*). Underneath the grave problem of worker safety that undermines the legitimacy of a communist party still claiming to represent workers, the disaster occurred almost six years after TAR government closed all gold mines in TAR, precisely under the pretext of protecting the environment (Li & Lin). And, of course, this disaster happened in Tibet, where anything of media worthiness could quickly escalate into political controversy with international repercussions. The combination of possible corruption, a fatal "accident," ecological crisis, and the *topos* of Tibet all added to the volume of the "silence."

It would be wise for global capital and local powers to keep quiet as well, so that they can continue to make fortunes from the "invisible"

Tibet. Hubris of capital doesn't seem to like prudence, however, even for tactical reasons. Central China Goldfields (CCG), a British miner already established in Tibet before the 2007 policy change, continued to operate there despite the Dalai Lama's appeal to Western mining companies to reconsider "ethical values." Evoking the Chinese name for "Tibet," *Xizang,* which means "Western treasures," a reporter from *The Telegraph* aptly calls China "the new El Dorado for years to come" (*Sunday Telegraph*). El Dorado, the legendary "lost city of gold," is now rediscovered in "Western Treasures" of the People's Republic. Who would want to have the golden goose killed by a mere "natural disaster?"

D. ECO-POLITICAL ZONES OF ENVIRONMENTAL GRIEVANCES AND PROTESTS

The picture about the "first-world" China tells a vastly different story. In the past decade and half, environmental grievances and protests have spread over China proper, and the trend has accelerated since 2007. A vice-chair of the Chinese Society for Environmental Sciences reveals the fact that "the number of environmental protests has increased by an average of 29 percent every year since 1996, while in 2011 the number of major environmental incidents rose 120 percent" (qtd. in Tobias). Another public figure has gone further to indicate that "[p]ollution has replaced land disputes as the main cause of social unrest" (qtd. in *Bloomberg News*). This growing volume of public protest in the more developed regions of China places in sharp contrast the silenced Tibet and marks an increasing division of the PRC into an urban "middle-class" North and the rural, peripheral South.

China observers have long expected her ecological crisis to become a major driver for political change and participatory governance. Environmental issues, it is presumed, have indiscriminative and universal impact upon the whole population. As long as the protest is not overly politicized to threaten Communist rule, public demonstrations can be managed under the rubric of *zhili,* a recent Chinese rendering of "governance" with more managerial-bureaucratic coloring. As a result, authorities respond to environmental protests in a manner more benign than they do to other "mass incidents" of social unrest. Provincial and local governments tend to make compromises or even cancel polluting projects outright to keep the public anger from spilling over to more politically sensitive arenas.

Up in the air, protests against choking air pollution in urban areas and industrial centers receive more official tolerance than its fair share. There is no way the rising economic superpower can import fresh air from overseas markets as it does for all other commodities—including farm land and fresh water in the form of its growing grain imports. On the contrary, China's air pollution has a broad international spillover effect. It clouds the sky over the Korean peninsula and Japan, dropping acid rain on its East Asian neighbors, long before some pollutants crossed the Pacific Ocean to reach the west coast of North America.

Domestically, the rising middle class has embarked on a tug of guerrilla war with authorities through social media and inflicted a lot of damage to the credibility of the CCP rule, even though their conspicuous consumption is a major contributor to the thick smog over Chinese cities. More rights-conscious and NIMBY ("not in my backyard"), China's urban middle class is becoming much like their counterparts elsewhere in the world, and China's rulers have to treat them as a long-term stabilizing force and give their grievances open yet manageable outlets.

A series of public demonstrations against paraxylene (PX) projects is a case in point. Paraxylene is a carcinogenic petrochemical with broad industrial application. China is dependent on import for forty percent of its PX needs. Capital intensive and strategic in the supply chain of international petrochemical industry, PX manufacturing can be highly profitable and provide a steady stream of tax revenue for provincial and local governments. Highly toxic yet highly profitable, PX projects also create a huge room for official rent-seeking in an already opaque regulatory environment. And it is the combination of toxic air and underhanded corruption that gives rise to public suspicion and protest.

Since 2007, three proposed PX projects in the port cities of Xiamen, Dalian, and Ningbo have been canceled or forced to relocate after citizens in these cities took to the street. Of the three, the demonstration in Ningbo went violent and caused much damage; having failed to control the violence the city government was forced to close the planned project while continuing to lament the loss of revenue and jobs. On May 7, 2013, a refinery and PX project proposed for a site just forty kilometres from Kunming, the capital of Yunnan province that borders Myanmar, Laos, and Vietnam, attracted thousands of protesting residents in a public demonstration marked by heavy police presence. The mayor of Kunming responded by promising a further "feasibility study" and following a "democratic process to consult public opinions" (Zhang). Distrust of the official promise and the lack of transparency, however, led to a second demonstration on May 16. The mayor again reassured the public at the scene that "whether the PX project goes ahead would be at the residents' discretion in July" upon the completion of the feasibility study and public consultation" (Chang).

In contrast to the silence of national media outlets towards previous anti-PX demonstrations, the newly installed leaders in Beijing apparently got it right where their own interest lies. Just one day after the first Kunming demonstration, the *People's Daily*, the flagship paper of the Communist Party, published an editorial calling for open dialogue between local governments, developers, and the public. It places the burden of proof squarely on the local government. More significantly, the editorial connects the PX-related public protests since 2007 to a broad agenda of "rebuilding national consensus for reforms," acknowledging that the previous "consensus" established in the Deng Xiaoping era of 1980s and 1990s has been "ruptured." In the end, the editorial invokes the new CCP leader and president of China: "Comrade Xi Jinping once pointed out that the Party must tear down the wall that insulates it from the masses of the people. Dialogue and positive interaction between the government and the public on an equal footing is essential to the leveling of such an invisible wall" (*Caixin*). According to Ma Jun, a Beijing-based environmentalist and founder of the Institute of Public and Environmental Affairs, "Environmental issues are much higher on the agenda with the current leadership" (qtd. in *Bloomberg News*).

In China's fragmented regulatory environment, it remains to be seen if provincial and local authorities will actually change their plan. In fact, the Kunming refinery and PX project are considered strategic for China's energy security. The proposed complex stands at the terminal of the oil pipeline connecting it through Myanmar to the Indian Ocean, thus bypassing the Strait of Malacca and diversifying China's oil transport routes. In addition, the land-locked mountainous Yunnan province now must import much of its liquid energy from coastal China. The refinery and the PX project would significantly cut the costs and reduce its dependency. It may well turn out in the end that the provincial and local governments, allied with the powerful state-owned China National Petroleum Corporation, seeks political support in Beijing to continue the proposed project, while making some marginal changes to pacify the public. Whatever outcome there will be, a more open and participatory process is taking shape in China around high-profile petrochemical projects.

On the ground, water pollution and solid waste contamination surrounding the major cities are issue drivers in national eco-politics as well, while the same problems in the vast countryside are more localized, often to the degree of national invisibility. The Qidong protest is a good example for the former trend. On July 28, 2012, citizens took to the street to demand that the city government stop the construction of a waste water pipeline. The pipeline, which would have dumped industrial waste water into the sea next to the estuary of the Yangtze River near Shanghai, was to be part of a paper factory owned jointly by a Japanese company. An initially peaceful demonstration went out of control when gangsters allegedly joined the procession. About one thousand people stormed the city hall, stripped the major in an attempt to enrobe him with a protestor's T-shirt. Some protestors uncovered and displayed such evidence of official corruption and decadence as expensive imported wine and condoms. The title of an online report puts it well: "Qidong NIMBY protest that occupied the local government and stripped a mayor may mark a new era of grassroots activism in China" (Alia).

The success stories (if by "success" we mean the failure of local governments and developers in carrying out planned projects), however, are limited to urban high-profile protests. There is a differentiated pattern in governmental responses, determined by the public health effects of the project, the relative voice volume of demonstrators, and their methods of protest on the broad spectrum of political and social negotiation. This more differentiated pattern can be drawn up as a bifurcated eco-politics between urban, largely middle class-led activism and rural invisibility and silence. And the ecological crisis on the Tibetan Plateau falls into the latter category, as part of China's internal south.

The Plateau remains largely rural and has managed to maintain the best air and water quality relative to the rest of the PRC. Historically, the primary challenge to the ecosystems on the Plateau came from deforestation, then hydropower damming, and lately mining and associated soil erosion, water and solid waste contamination. Deforestation reached its barren peak in the 1990s before it was forcefully stopped in 1998, when a once-in-a-century flood along the Yangtze River threatened to submerge much of the Chinese heartland. While a massive number of trees has since been planted in the watershed regions of major rivers, some tributaries of the same rivers are running dry due to overuse or serial dam construction. Being the "most 'dammed' country" already (Chellaney), with many rivers running dry seasonally or permanently, China continues to implement a plan to raise the share of non-fossil fuels in its energy mix to 15 percent by 2020, up from 9.4 percent in 2011. Hydropower is expected to make the biggest contribution. And the Tibetan Plateau takes the brunt of damming because hydroelectric capacity is concentrated on the upper reaches of China's two major waterways, the Yangtze River and the Yellow River and their tributaries. These two waterway systems flow down the Plateau through the Tibetan regions of the Kham and Amdo, in today's Sichuan, Yunnan, and Qinghai provinces. So far as there is enough water of relatively decent quality to be diverted to the urban centers and coastal regions of the PRC, the NIMBY middle class, who are concentrated in these areas, won't make much a fuss, leaving these "development" problems to the technocratic elite.

The underground story remains buried underground or at best "covered" in the media and public forum as industrial "accidents." With coal burning making up nearly seventy percent of China's total energy consumption, the rate of death to coal output still remains ten times higher than the developed world (*China Labor Bulletin*). Problems of groundwater pollution, depleting aquifers, or even sinking landmass due to mining, are all part of the "underground" crisis, limited to experts and technocrats for solutions. Out of sight (or smell), out of mind—this NIMBY logic of urban environmental protests challenges the ideology of environmental universalism we discussed previously.

E. REVIVING THE SPIRIT OF TIBETAN INDIGENOUS RESISTANCE

Tibet as a *topos* can and should be reconfigured for an eco-political cartography to revive the spirit of her indigenous resistance against the presently silenced state of invisibility. This can be anchored on a dialectic of history and utopia, drawing from the successful experience of grass-roots activism as well as the established symbolic and cultural capital. There is the Shangri-La mythology that appeals to a utopian universalism for rebalancing human needs and natural provisions in an age of mass consumerism, as much as it used to be a political rallying point for the Tibet Cause. Inside the People's Republic, the Shangri-La myth and Shangri-La tourism development are entwined in a paradox of utopia and commodification, and continue to serve as source of inspiration to many Chinese seeking a renewed imagination of human-nature harmony already lost in their own ravaged homeland. Practically speaking, the millions of Chinese tourists may add more environmental stress to the Plateau, but most of them leave Tibet refreshed to develop a more empathetic understanding of Tibetan way of life. There are tens of thousands of Chinese volunteers, eco-tourists, and Buddhist pilgrims that engage Tibet annually and are devoted admirers of Tibetan cultural heritage. Despite the fact that the Shangri-La myth was created and developed first by Westerners and later incorporated into Chinese cultural discourse, the Shangri-La imaginary has the potential power of mobilizing more support from the vocal and active sectors of the Chinese population for Tibet's eco-political struggle.

The history of Tibetan indigenous resistance on the environmentalist front is less known to the outside world. I am not referring to the Tibetan Buddhist worldview and everyday practice that constitutes the traditional Tibetan way of life. The ideals are already incorporated into the Shangri-La discourse, can inspire other cultures, but the traditional way of life cannot be directly replicated elsewhere due to the unique eco-cultural environment of the Plateau. Rather, I will recount the story of the first grass-roots activism in the eco-political struggle, one that later fired and shaped a nationwide activism and helped institutionalize their endeavors in the establishment of a series of nature reserves on the Plateau. Equally illuminating is the absence of a modern, largely Western discourse of environmental ethics and eco-politics in the indigenous phase of the struggle—that is, the kind of environmentalism that emerged in the 1950s against extractive capitalism and industrial pollution in the global North. In that sense, the following story surrounding Kekexili should be viewed as the beginning of the eco-political south in the PRC. Thanks to the political-economic entanglement with the international production and consumption of shathoosh, the Kekexili struggle belongs to the Global South.

The story began with a gold rush that preceded the Gyama valley disaster by almost three decades. In 1984, gold was discovered in Kekexili (Mongolian: Hoh Xil; Tibetan: A-Chen Khunu), an uninhabited area about 50,000 square kilometres in southwest Qinghai province. Tens of thousands poor peasants from eastern Qinghai were recruited to sift gravel for the precious metal. Provincial authorities attempted to step in and began to grant mining permits up to thirty thousands in a controlled area of 450 square kilometres in early 1989 (see Schaller 77–78). This was simply a desperate move for the authorities to create some semblance of control and extract revenue from selling the permits. There was just no material and institutional infrastructure for law enforcement in the no-man's land. In May 1989 a blizzard hit the area and stranded thirty thousand miners—a storm that arrived too late in the season well after the annual rush got underway. The news of stranded miners hit the national media about the same time Beijing sent in the military for rescue. The name Kekexili thus began to enter into the national parlance of China.

About the same time our protagonist also entered the scene. A tall and hot-tempered Khampa warrior of the new era, Jeesang Sonam Dhargye (1954–1994), turned himself into a dedicated guardian of wildlife from a firm believer in modern technocracy and a prospective gold explorer. During the Cultural Revolution he graduated from a three-year program at the Qinghai Institute of Nationalities in 1974—a privilege reserved for young people of correct, "proletarian" background. Upon graduation he could have stayed in the provincial capital, Xining, or even gone to work in Beijing. But he went back to his hometown and chose a middle-school teaching career instead, the least glamorous profession for a college graduate in the anti-intellectual environment of the time. In 1985, when Sonam was working as the deputy head of the county school administration, a devastating blizzard destroyed local nomads' livestock and lifetime savings. He participated in the rescue efforts and witnessed the harsh realities of the fragile subsistence economy subject to the whims of unforgiving nature. This prompted him to change his career path and sought an official job transfer to become the Party chief of Suojia district, Zhiduo (Dritoe) county in Yushu Tibetan Autonomous Prefecture of Qinghai.

Sonam's ideological background does not fit into a standard profile of a Tibetan environmentalist. "An atheist and nonbeliever in Buddhism, he considered the pilgrims on their way to Lhasa unproductive and wasteful. The lamas, in his eyes, have everything supplied and handy, lived like kings, and are 'eaters of human flesh'" (Liu 88). From 1987 to 1989 Sonam spent two summers with a geological survey team in his home district and learned all the tricks of geological positioning and mapping and identifying and screening various mineral ores. The geologists even worked with him to locate a hydropower dam site. They alerted him to the ongoing gold rush in Kekexili and advised him to make a quick buck for his poverty-stricken district. "Now that you know very well that animal husbandry will lead you nowhere for modernizing the local economy, why not go west to Kekexili, for gold?" (Liu 90).

To their surprise, Sonam knew little about Kekexili, even though the uninhabited region was under the jurisdiction of his district. The local nomadic population of about six thousands were concentrated to its eastern part, separated from Kekexili by mountains and marshes on the very upper reaches of the Yangtze river, called *Tongtianhe* in Chinese or "The River that Leads to Heaven." By 1992 he finally persuaded the prefectural Party chief to establish the Western Work Committee of Zhiduo County CCP Committee, himself serving as the deputy chief of the county committee in charge of the work committee. Thereafter he established a company, named it "Zhiduo County Economic and Technological Development Co." and appointed himself as its general manager. "I'm the CEO of a company with the largest territory in the world, in charge of more than 50,000 square kilometres. But I'm also the poorest one, without a single penny in the book," Sonam joked with his confidant (Liu 91).

On his first trip to Kekexili Sonam brought with him a handbook of industrial minerals. On his last trip, that book vanished altogether with the drive for gold and was replaced by the loose leafs from a directory of endangered species. The catalyst for such a conversion was not religious epiphany or sophisticated philosophizing. When Sonam and other members of the work committee stepped on the edge of Kekexili for the first time, they noted truck tracks crisscross the scarred landscape of gold extraction and, alas, thousands of skinned corpses of chiru—the famed Tibetan antelope known as the source of shahtoosh. In the early 1990s, the gold rush receded in Kekixili, only to the rise of chiru poaching.

According to George Schaller, American naturalist and conservation scientist, as late as 1985 the slaughter of chiru had yet to reach Kekexili. The first sight he had of skinned chiru was in 1988. This coincided with the growth of the shahtoosh shawls as "a fashion statement of the world wealthy" in the late 1980s (see Schaller 8 and ch. 8, "Deadly Fashion"). It is by that time the opening of trade and marketization also began to penetrate into the far west of China. International demand for shahtoosh grew rapidly and the chiru population declined accordingly. Traditional trade linked the Tibetan Plateau to the nearby Indian state of Jammu and Kashmir, the production center of shahtoosh shawls and scarfs. The producers had little knowledge about the source of their raw materials or were led to believe the traders' fiction that that it was plucked from the bushes and rocks (Gupta 61).[3] Farther removed geographically and socioeconomically, the wealthy consumers could care even less, even if shatoosh trade was banned since 1979. For all they knew, "[t]he source of the wool is the Mountain Ibex of Tibet. After the arduous Himalaya winter is over, the Ibex sheds its down undercoat by scratching itself against low trees and bushes," so advertised by an upscale department store in New York in the early 1990s (qtd. in Schaller 68).

The global network of fashion production and consumption took a heavy toll on the most vulnerable species in a most remote corner of the world. And only their human guardian bore witness to the sheer brutality of their massacre. Soman and his small team, called Wild Yak Brigade, had no choice—and needed no eco-philosophical inspiration for their action—but to follow their own conscience. The "wild yaks" simply followed their "Tibetan instinct" to defend the lives of their wildlife companion in spite of their leader's pronounced disdain of the ecclesiastic hierarchy. Vastly underfunded and outnumbered, his team managed to conducted twelve patrols of Kekexili and captured poachers on eight occasions in the first year. On January 18, 1994, he and three other team members came across eighteen poachers carrying over 1,800 chiru skins. A gun battle ensued, and Sonam Dhargye was killed in action. When reinforcement arrived a few days later, he was found frozen, still maintaining his last fighting posture.

"To accomplish something [of the mission the wild yak brigade was formed to carry out] in China," Sonam remarked before his death, "some people must die [to create a crisis and alarm the public]" (Liu 100). He was cremated, the highest honor normally reserved for a prominent reincarnate lama. A well-respected reincarnate lama presided over his funeral. This lama was once quoted as saying "a true Communist Party member is as good as a *Bhikkhu*." Indeed, the funeral was held in the rituals blending Buddhism and Communism alike. Sonam's body was covered with a flag of the Chinese Communist Party and guarded by a paramilitary soldier on each side, which are all encircled by a thousand butter lamps (Liu 248–49).

Soon after his death the Wild Yak Brigade regrouped and expanded to have more than fifty volunteer members, most of whom were Tibetan. Meanwhile, the story of his sacrifice and martyrdom quickly spread, and a series of "first" ensued. A year after his death, the National Bureau of Environmental Protection (NBEP) and the Ministry of Forestry named him "National Guardian of the Environment," the first such honor ever extended to any individual citizen of the PRC. Chinese volunteers joined force. Yang Xin, an environmentalist photographer and author who once successfully rafted on the Yangtze River from its source region in Yushu all the way to Shanghai, led a nationwide campaign of fundraising and publicity. Yang enlisted the support from the highest official in Qinghai province and the NBEP as well as leading scientists and journalists working on the environment. In 1996 the construction of a nature reserve station— named after Sonam Dhargye—began near the entry point on the highway to the area of his last battle. The structure was built on permafrost near 5,000 km above the sea level. Its completion in 1997 marked the first such facility in China entirely funded by public donation and built by volunteers. In 1998 a former team member of the Wild Yak Brigade joined two of Sonam's former students in forming the first NGO ever established by Tibetans inside the PRC, and named it Eco-Economic Development Council of Pan-Yangtze River Source Region (Yang).

Concerned parties from outside China also joined force. In 1997, International Fund for Animal Welfare (IFAW) became the first international organization to fund the patrol with a donation of USD 10,000 (Morton). George Schaller's *Wildlife of the Tibetan Steppe,* published in 1998 on the endangered species of Tibet, further galvanized the campaign. In October 1999 an international workshop was held in Xining, the capital of Qinghai. At its conclusion the Xining Declaration on the Conservation of and Control of Trade in Tibetan Antelope was adopted. It calls for all governments to strengthen law enforcement against shahtoosh trade. Sponsors of the Declaration include government representatives from China, France, India, Italy, Nepal, the UK, and the US, as well as representatives of almost all major international and Chinese organizations in wildlife protection, such as the Convention on International Trade in Endangered Species of Wild Fauna and Flora (CITES), China Wildlife Conservation Association (CWCA), International Fund for Animal Welfare (IFAW), Tibetan Plateau Project (TPP), TRAFFIC (the wildlife trade monitoring network), Wildlife Conservation Society (WCS), Wildlife Protection Society of India (WPSI) and World Wide Fund for Nature (WWF). (*Xining Declaration*)

In 2004 the feature film *Kekexili* was released, adding to the corpus of publications and documentaries on the Sonam Dhargye legend (see Chen for more information about the film). Thanks to the film's popularity in China and success at international film festivals, the term "Kekexili" became almost a household name in China and, in some circles, a fashion statement for youth idealism. Whenever I give a lecture or teach a college class in China, I ask the audience to identify Shangri-La and Kekexili. Many know the former as a nice tourist site in Yunnan province but little else in connection to its Western origin and imagination of Tibet. Almost all people identify Kekexili with the Plateau and Tibetan antelope; some even shout out the hero's name, "Sonam Dhargye!"

F. EPILOGUE

The struggles around Kekexili and associated bureaucratic hurdles and funding headaches, not to mention the everyday risks and continued violence in a virtually lawless environment, are far more tortuous than my brief chronicle would suggest.[4] Indeed, there is always the risk of trending towards hagiographical, larger-than-life heroes. Ideological appropriation of messy, real-life experiences into simple and simplistic representation is almost the rule of the game. Richard Harris observes that Sonam Dhargye actually displayed what he calls "the prevailing Chinese instrumentalist altitude towards wildlife." Commenting on a documentary about Sonam, Harris jibes, "Whether or not wildlife conservation is central to Soman's martyrdom, it was clearly viewed by the filmmakers as a weak rationale for idolizing their protagonist. Instead, the viewers were invited to regard Sonam almost as an updated, ethnically Tibetan Lei Feng, doing good deeds out of the goodness of his heart (which, if during the 1950s had meant selflessly serving the people, by the 1990s had come to mean advancing economic development)" (61). Harris is correct (and seems to be somewhat dismayed by the fact) that Sonam and members of the Wild Yak Brigade were not practitioners of Western Deep Ecology or traditional Buddhist pacifism. The historical and theoretical significance of their struggle lies precisely elsewhere: the contingency of history and the equally contingent, open-ended process of subject formation for social agency. To expand and play on the dated Marxian tenet for our purpose, human nature is the total sum of human-nature relations. Transcendence of historically-conditioned and ecopolitically-situated contradictions remain a luxury for arm-chair "interventionists" or, shall we say, the Dalai Lama.

There are many questions that remain to be answered—some of which may not have an answer. With the series of "first" bestowed upon the hero and those he has inspired in China, how can we confront the deep-seated paradox between the valorization of an ethnic Tibetan *and* national hero of the PRC, and the present state of silence on the Plateau? To flip the coin and take a look at the other side, how could the poachers, mostly of Muslims of the Hui minority from eastern Qinghai, put their financing and armed operations together in a tightly controlled, authoritarian and communist state? What drove them into such a desperate action? Outside the PRC, the chiru trade connects the Plateau to the state of Jammu and Kashmir, with transit routes passing the Himalayas through Nepal or other regions of India. What impact has the law enforcement against the trade left upon the livelihood of the shahtoosh producers? What have the well-intended international organizations of wildlife protection done to ease the disruption that their action may have inflicted upon the local economy? Or do international organizations have the capacity and will to help solve the problems of the Global South that are steeped in inequality of the human world, which should be primarily defined between the North and South in the era of transnational capital, as well as inequality between humans and nature?

When I began to look into the Kekexili resistance and the inspiration that the Wild Yak Brigade has drawn to China, two unrelated things crossed my mind, almost by accident. About a century ago, Vladimir Lenin theorized—and devised a strategy to implement—a roadmap towards the Bolshevik revolution (and I paraphrase it here): the revolution is more likely to take place and succeed at the weakest link in the international system of capitalism. That meant Russia and, three decades later, China. On the global eco-political front, historical events seem to have unfolded more concurrently than sequentially. About the same time the two landmark technocratic documents were inked and announced as the Stockholm Declaration and The Club of Rome Report, the brave women (and some men) on the Indian side of the Himalayas started what would later become the Chipko Movement. Today arm-chair eco-philosophers and eco-feminists draw inspiration from the original "tree huggers," do they still remember that these prototypical native "eco-feminists" started their nonviolent resistance to preserve their basic livelihood? Just as Sonam Dhargye's "Chinese" attitude towards wildlife is "instrumentalist," these Indian women's approach to firewood, fodder, and other sustaining materials for their collective survival cannot but be "instrumentalist", too. Ecological and economic exploitation in the Global South cannot and should not be separated, from the Himalayan villagers of the Chipko Movement to the Wild Yak Brigade, and to the Gyama valley landslide that began with our investigation.

WORKS CITED

Alia. "Qidong NIMBY protest that occupied the local government and stripped a mayor may mark a new era of grassroots activism in China." *Offbeat China*. 28 July 2012. Web. 14 Aug. 2013.

Bloomberg News. "China Leaders Signal Tolerance for Slowdown." Bloomberg News. 26 May 2013. Web. 14 Aug. 2013.

Caixin. "The People's Daily Editorial: A Path towards Resolving Public Protests against PX Projects" (in Chinese). Caixin Media. 8 May 2013. Web. 14 Aug. 2013.

Chang Meng. "Thousands protest Kunming PX plan." *Global Times*. Global Times. 17 May 2013. Web. 14 Aug. 2013.

Chellaney, Brahma. "The World's Most 'Dammed' Country." *India Today*. India Today. 26 Dec. 2011. Web. 14 Aug. 2013.

Chen, Jie. *Nation, Ethnicity, and Cultural Strategies: Three Waves of Ethnic Representation in Post-1949 China*. Unpublished PhD dissertation. Rutgers University, 2008. Print.

China Labor Bulletin. "Report claims coal mine deaths in China fell by one third in 2012." China Labor Bulletin. 29 Jan. 2013. Web. 14 Aug. 2013.

China Merchant Daily. "The Family of a Shaanxi Worker Buried under the Landslide at a Tibet Gold Mine Receives One Million RMB for Compensation" (in Chinese). *sn-ifeng.com*. 8 May 2013. Web. 14 Aug. 2013.

"Corporate Overview." *China Gold International Resources Corp*. Web. 14 Aug. 2013.

CNW. "China Gold International Reports Strong 2013 First Quarter Results and Reconfirms 2013 Annual Outlook." CNW, a PR Newswire Company. 15 May 2013. Web. 14 Aug. 2013.

Environment & Development Desk (EDD), Department of Information and International Relations, Central Tibetan Administration. "Assessment Report of the Recent Landslide Event in the Gyama Valley: Its Possible Cause and Impacts." *Tibet.net*. 9 Apr. 2013. Web. 14 Aug. 2013.

Gupta, Saloni. *Contesting Conservation: Shahtoosh Trade and Forest Management in Jammu and Kashmir, India*. Unpublished PhD thesis. School of Oriental and African Studies, University of London. 2012. Print.

Harris, Richard. *Wildlife Conservation in China: Preserving the Habitat of China's Wild West*. New York: E.M Sharpe, 2007. Print.

Huang, Hongxiang. "The Silence Around Tibet's Ecological Crisis." *The Atlantic*. The Atlantic Monthly. 7 May 2013. Web. 14 Aug. 2013.

Lin, Hongmei and Li Zhongfa. "Tibet [TAR] Closes All Gold Mines to Become the Best Protected Region of the World" (in Chinese). *ChinaNet*. 18 July 2007. Web. 14 Aug. 2013.

Liu Jianqiang. *Heavenly Pearl, A Tibetan Legend* (in Chinese). Lhasa: Tibet People's Publishing House, 2009. Print.

Morton, Katherine. "Transnational Advocacy at the Grassroots: Benefits and Risks of International Cooperation." Eds. Peter Ho and Richard Louis Edmonds. *China's Embedded Activism: Opportunities and Constraints of a Social Movement*. London & New York: Rouledge, 2007. 195–215. Print.

Pomeranz, Kenneth. "The Great Himalayan Watershed: Agrarian Crisis, Mega-Dams and the Environment." *New Left Review* 58 (July–Aug. 2009): 5–39. Print.

Richards, Thomas. "Archive and Utopia." *Representation* 37 (Winter 1992): 104–35. Print.

Schaller, George B. *Tibet Wild: A Naturalist's Journeys on the Roof of the World*. Washington,DC: Island Press, 2012. Print.

———. *Wildlife of the Tibetan Steppe*. Chicago: University of Chicago Press, 1998. Print.

Schell, Orville. *Searching for Shangri-La from the Himalayas to Hollywood*. New York: Metropolitan Books, 2000. Print.

Sun, Yanfei and Dingxin Zhao. "Environmental Campaigns." Ed. Kevin J O'Brien. *Popular Protest in China*. Harvard University Press, 2009. Print.

Sunday Telegraph. "China and Britain ready to exploit Tibet's natural resources." The Telegraph. 27 July 2008. Web. 14 Aug. 2013.

Tibet Huatailong Mining Development Co. Web. 14 Aug. 2013.

Tobias, Michael Charles. "China Declares Global State of Emergency: An Urgent Telegram from Taihu." *Forbes*. 21 May 2013. Web. 14 Aug. 2013.

Weise (Tsering Woeser). "Landslide Induced by Frenzied Mining at Hometown of Songtsen Gampo is said to be a 'Natural Disaster'" (in Chinese). Invisible Tibet, 30 Mar. 2013. Web. 14 Aug. 2013. Trans., High Peaks Pure Earth. High Peaks Pure Earth, 30 Mar. 2013. Web. 14 Aug. 2013.

Xining Declaration: On the Conservation of and Control of Trade in Tibetan Antelope. Xining, China. 12–14 Oct 1999. *cites.org*. Convention on International Trade in Endangered Species of Wild Fauna and Flora. Web. 14 Aug. 2013.

Yang Xin et al. *Experiencing Kekexili for a Decade: Volunteers Telling Their Own Stories"* (in Chinese). Beijing: Sanlian Books, 2005. Print.

Zhang Tao, "It All Depends upon Public Approval Whether or Not the PX Project Will Move Ahead, Says the Mayor of Kunming" (in Chinese). *Caixin*. 10 May 2013. Web. 14 Aug. 2013.

NOTES

1. Various media reports—limited as they were—were summarized in a document prepared by Tibetan exiles in India. See Environment & Development Desk (EDD).

2. See *CNW*. The EDD report gives more detailed information: "The key minerals produced are copper (Cu), gold (Au), silver (Ag), molybdenum (Mo), lead (Pb) and zinc (Zn). The contributions of these mineral ores to gross revenues are 75 percent, 11 percent, 7 percent and 6 percent. The mine is expected to get an annual production of 176 million pounds of copper, 35 thousand ounces of gold, 2.7 million ounces of silver and 2.3 thousand tons of molybdenum" (Environment & Development Desk 5).

3. For a though study of shatoosh trade and production, see Gupta.

4. See Liu chapters 3 and 6 for more details. For an English source, see Yanfei Sun and Dingxin Zhao, "Environmental Campaigns."

THREE

Red China, Green Amnesia

Locating Environmental Justice in Contemporary Chinese Literature

Cheng Li and Yanjun Liu

A. CHINESE ENVIRONMENTAL JUSTICE AS A RESEARCH AGENDA

When a shroud of haze rests over Beijing, the Chinese people choose to trust the $PM_{2.5}$ data reported by the U.S. embassy in Beijing to evaluate their air quality. Particles less than 2.5 micrometers in diameter ($PM_{2.5}$) are referred to as "fine" particles and are believed to be consistently and independently related to the largest health risks, including lung cancer and other cardiopulmonary mortality.[1] The Chinese government refused to report $PM_{2.5}$ until 2012. An index reading of $PM_{2.5}$ below about 50 is considered healthy. On December 4th, 2011, the U.S. embassy reading for Beijing reported a $PM_{2.5}$ concentration of 522 micrograms per cubic meter ($\mu g/m^3$) beyond 500$\mu g/m^3$, the top of the index, showing a hazardous air condition. In contrast, the Chinese environmental bureau report, which failed to take $PM_{2.5}$ into consideration, indicated only slight pollution on that day. As a later CNN news titled "China: Haze isn't foreign embassies' business" (Jiang) notes, confronted with the embarrassing, continuous and inconsistent environmental monitoring data provided by the U.S. embassy and Wu Xiaoqing, the Vice Minister of Chinese Environmental Protection Department, denounced the U.S. embassy's data as irresponsible as it was not in accordance with the Vienna Convention on Diplomatic Relations and the Vienna Convention on Consular Relations, which both state that diplomats should not interfere in the internal affairs of the receiving state (United Nations). By citing diplomatic rules as a rhetorical strategy, the Chinese government made it clear that it regarded a foreign embassy's information release on Chinese environmental conditions as a violation of sovereignty and as an interference in Chinese internal affairs. With Chinese environmental scandals increasing dramatically due to the country's extensive economic growth and the sharing of information online at unprecedented rates, Wu's defense appeared powerless.

Given that even Beijing, as the capital of China, faces such environmental quandaries, one can imagine how serious environmental degradation must be in other inland cities and rural regions of China, which are often rendered as deserted corners through selective amnesia, in sharp contrast to the mass environmental concern in big cities like Beijing with updated and relatively transparent environmental data, or in coastal cities like Xiamen and Dalian with active and powerful residents who protest against chemical plants in their communities. For instance, the officials in Shenqiu, a cancer village [*aizheng cun*] caused by toxic chemical pollution into the Huai River in Henan Province, China's most populated province, blame reporters who disclose the grotesque conditions of the abject farmers, some of whom have to use a catheter to remove the contaminated blood from their vulnerable bodies.[2] Shenqiu's officials do not deny the report. Instead, they argue that their village is not distinctive and insist that the journalists objectively cover all the other polluted neighboring villages as well.

There has been a wide range of scholarship on Chinese environmental justice. Some scholars tend to focus on nonfiction narratives of Chinese environmental justice. Adam Moser's "The Yangzhonghai Case: Struggling for Environmental Justice" (2013) argues that Chinese executive and judicial branches should play an active role in law enforcement to ensure the environmental justice of the poor. Judith Shapiro's *China's Environmental Challenges* (2012) provides an empirical analysis of environmental injustice by pointing out regional, national and transnational injustices. She also illustrates that the transnational impacts on China are double-folded: China serves as both an importer and exporter of environmental harm. Elizabeth Economy's *The River Runs Black* (2010) provides a historical dimension to the Chinese environmental predicament such as the cancer villages resulting from environmental degradation of the Huai River, the Chinese people's environmental consciousness, and various grass-roots environmental campaigns.

In spite of their non-Chinese focus, some western scholars' environmental justice studies are of relevance to issues of Chinese environmental justice. Ole Pedersen illustrates the specific justices under the umbrella of environmental justice, including distributive justice, procedural justice, recognition, all-encompassing protection, productive justice and ecological justice. He observes that in the Third World, "there is a feeling of injustice requiring an integrated approach that involves more than environmental issues, but includes social justice, food security, cultural recognition, equity and democratic participation. In many ways the experience of the poor nations echoes that of the poor minority communities within the United States" (19). Stacy Alaimo has coined the term "trans-corporeality," which deftly argues that the human body is "vulnerable to substances and flows of its environments, which may include industrial environments and their social/economic forces" (28). Rob Nixon's *Slow Violence and the Environmentalism of the Poor* (2011) examines the environmental trauma of disadvantaged and disenfranchised poor in a postcolonial context by coining "slow violence" as a delayed violence that "occurs gradually and out of sight" and "is dispersed across time and space" (2). Nixon specifically refers to climate change, toxic drift, deforestation, oil spills, and other environmental threats. Lawrence Buell labels "toxic discourse" as "expressed anxiety arising from perceived threat of environmental hazard" (*Writing* 31). Buell contends that "the threat of hegemonic oppression is key to toxic discourse" (*Writing* 41). Ursula K. Heise's *Sense of Place and Sense of Planet* (2008) argues that ecocosmopolitanism might invite a sense of planet that arises from global shared risks to encourage local environmental justice movements to participate in a broader sense of culture.

Studies have also been published that touch upon environmental justice in contemporary Chinese literature. Yue Gang's "Fragments of Shangri-La" (2010) discusses Tibet's paradoxical colonial and independent status that has led to its environmental exploitation. Jerry Varsava examines ecological and ethnic deterioration in Inner Mongolia brought about primarily by the Han Chinese intrusions—this theme is further explored in Jiang Rong's novel *Wolf Totem*. Karen Laura Thornber's thoughtful observations and nuanced insights into "ecoambiguity" prevalent in literature from East Asia are groundbreaking as well as challenging, considering this region's long cultural history celebrating the intimate ties between humans and nature even as its people damaged environments. According to Thornber, ecoambiguity includes:

> ambivalent attitudes toward nature; confusion about the actual condition of the nonhuman, often a consequence of ambiguous information; contradictory human behaviors toward ecosystems; and discrepancies among attitudes, conditions, and behaviors that lead to actively downplaying and acquiescing to nonhuman degradation, as well as to inadvertently harming the very environments one is attempting to protect. (6)

Thornber considers environmental ambiguity in Jiang Rong's novel *Wolf Totem* as something between attitudes and behaviors, attitudes and conditions, and rising population and limited resources (314–18).

To a large extent, current Chinese environmental literature study is still first-wave oriented, bearing a preservationist view. Most native Chinese ecocritics tend to ignore environmental justice concerns.Yang Jincai observes that "the major criterion for judging a piece of literature is to see whether its author holds an ecological stance . . . devoid of human subjective dominance or anthropocentrism" (195). Wang Nuo maintains the vital role of a biocentric viewpoint in ecocriticism, which influences many Chinese ecocritics. His latest book *Shengtai Piping Yu Shengtai Sixiang* (*Ecocriticism and Ecologism*, 2013) insists on the importance of biocentric notions. Other Chinese ecocritics, including Lu Shuyuan, Cheng Hong, and Wei Qingqi, tend to be first-wave oriented.[3] Wang Nuo and other Chinese ecocritics influence research conducted by younger ecocritics in China, including the ecocritical studies on contemporary Chinese literature: *Shengtai Yishi YuZhongguo Dangdai Wenxue* (*Ecological Consciousness and Contempoary Chinese Literature*, 2008) written by Wang Shudong and *Shengtai Piping Shiyuxia De Zhongguo Xiandangdai Wenxue* (*An Ecocritical Perspective into Modern and Contemporary Chinese Literature*, 2009) edited by Wang Xirong. Wang Shudong and Wang Xihua examine contemporary Chinese writers, including Chi Zijian, Zhang Wei, Wei An, Zhang Chengzhi, and Li Cunbao.

Therefore, to promote the transformation of Chinese environmental criticism, it is worthwhile to ask the following questions: Is the Beijing $PM_{2.5}$ information released by the U.S. embassy on Chinese soil a new form of "green imperialism" as the U.S. hegemony is keen on setting environmental standards for the rest world and controlling the discourse related to environmental issues? How can the environmental concerns of Chinese subalterns—namely, farmers and ethnic minorities—be justly represented in public sphere? Specifically, given that the second-wave ecocriticism underlines the environmental justice of non-white, female and homosexual concerns in Anglophone areas, how can we best approach the issue of environmental justice invoked by contemporary Sinophone writers with regard to either their representation or epistemology?

In this article, we argue that Chinese environmental justice occurs at three dimensions: transnationalism, class and ethnicity. In the event of Beijing's $PM_{2.5}$ report, the Chinese officials awkwardly use the concept of sovereignty geographically identified by territorial boundaries as a rhetorical strategy to deny Chinese people's access to international information, with the objective of hiding the Chinese government's tardiness and impotence in coping with the environmental crisis. In the vast rural areas of China, most farmers still cannot freely, legally and permanently leave the lands of their fathers even when the lands have been badly contaminated and even when they finally get a temporary job in remote cities. These unfortunate Chinese farmers have to silently suffer from the unfair distribution of environmental public goods due to political and economic disempowerment. In the ethnic minority areas, the Tibetan pure and holy land has been rudely intruded by the Han Chinese, who make up 92 percent of China's population and have persistently carried on socialist transformation to civilize the supposed barbarian Tibetan since 1951. Consequently, the Tibetan distinctive ethnic perceptions and practices relevant to natural environment have been in serious decline. In short, transnationalism, class and ethnicity are three dimensions essential to an understanding of the status of environmental justice in contemporary Chinese literature.

We will first briefly reexamine the transnational implications of the Beijing $PM_{2.5}$ case from an updated postcolonial perspective and then turn to the main body of this article where environmental criticism is applied to three Chinese literature texts: Yan Lianke's *Streams of Light and Time* (2009 [1998]), Ma Jian's *The Dark Road* (2013), and A Lai's *Empty Mountain* (2005).

Yan Lianke's *Streams of Light and Time* investigates an invisible and enduring toxicity impacting Chinese farmers who cannot live beyond forty years in their home villages and are not allowed to migrate due to Chinese migration law. Unfortunately, even though four village leaders throughout forty years strived to live longer in various ways, their efforts are all futile. Ma's novel, *The Dark Road* offers another prospect: what will happen if the trapped farmers try to challenge the migration law. The fugitive road turns out to be an odyssey by waterway undertaken by the pregnant Meili and her husband Kongzi with their young daughter to escape the atrocious one-child policy. After a painful and long journey, they finally come to a place where one-child policy is not enforced strictly because the polluted environment is so hostile to reproduction. A Lai's *Empty Mountain* moves Yan and Ma's discussion about trapped Han Chinese farmers to Tibet. Tibetan natural resources, together with the linguistic and religious culture of Tibet, have been severely abused by the Han Chinese.

By textual analysis, we argue that the marginalized rural farmers and displaced ethnic minorities can be used as entry points to approach the issue of environmental justice in Chinese literature despite the fact that the current visible Chinese environmental campaigns are mostly launched for the rights of the urban population. These three Chinese novelists have unintentionally envisioned the emergence of environmental justice issues from the new territories of Chinese environmental literature studies through their texts. As an attempt to orient the so-called second-wave Chinese ecocritical studies, we call for more attention to the voice of the underprivileged and the "slow violence" in contemporary Chinese environmental literature studies.

The three Chinese novelists are not usually judged ecocritically either by Chinese or Western literature scholars even though they have implicitly or explicitly demonstrated environmental justice concerns. As Karen Thornber argues, "ideally in the years to come scholars of non-Western literatures will publish more work with an ecological focus and more ecocritics with a background in Anglo-American literatures will take up study of non-Western languages" (452). By analyzing the environmental justice discourse in these works of Chinese literature, we hope to shed some new light on the intersection of East Asian studies, ecocriticism, and comparative literature.

B. IN THE SHADOW OF TRANSNATIONAL ENVIRONMENTAL JUSTICE: DISTORTED GREEN IMPERIALISM

By coining the phrase "green imperialism," Richard Grove examines the deleterious effects of deforestation, drought and soil erosion, resulting from crass exploitation and the destruction of traditional balance between populations and resources by European worldwide colonial expansion. In the event of Beijing's $PM_{2.5}$ report, however, we can see how the Chinese officials implicitly use the discourse of "green imperialism" to recall historical western colonial impacts, especially on the environment in the Global South, thus discrediting the environmental monitoring information released by western authorities which is becoming more and more accessible to ordinary people in China. Vice Minister Wu Xiaoqing's assumption that environmental sovereignty should be free of foreign intervention betrays a looming nationalism that is toned with postcolonial discourses. After the 1989 Tiananmen Square incident, nationalism has been lastingly utilized in Chinese literature to assert the necessity of authoritarian power domestically as a shield against the foreign intervention of internal affairs. Politically, whenever democratic process is put forward, the Chinese government always accuses the western countries of intervening in its sovereignty. Economically, qualified Japanese goods are boycotted as Japanese officials deny the invasion of China during WWII, which "badly hurts the feelings of the Chinese people." Adopting a "strategy of essentialism" to name people regardless of ages, ethnic minority, class and income as Chinese, the officials deliberately neglect the disenfranchised and poor communities that are distinct from those in power.[4] More importantly, postcolonial discourse has been morphed into a nationalized stance to justify the official acts of oppression which deny citizens' access to environmental information from western countries. Thus, it is necessary to examine environmental nationalism through postcolonial lenses.

In *Orientalism* (1979), Edward Said examines notions of Orientalism since the 18th century by exposing the power of discourse which facilitates and upholds colonialism in oriental regions. He later notes that "the power to narrate, or to block other narratives from forming and emerging, is very important to culture and imperialism" (*Culture* xiii). Most Chinese scholars seem to think that Said and his followers only focus on the impacts of colonial history, especially cultural impacts on oriental countries colonized by western hegemonic powers. Therefore, little attention is given to the "oppositional criticism" embedded in the postcolonial (also postmodern) theory to elucidate the broadly oppressed.[5] Said asserts that "the criticism must think of itself as life-enhancing and constitutively opposed to every form of tyranny, domination, and abuse" (*The World* 29). "Tyranny, domination and abuse," these traditional thematic concerns of the colonized territories, are also the contested issues in a broader sense, covering all powerless and oppressed peoples. In other words, Said's resistance against the hegemonic powers is not restricted in the narrow range of the western countries and their corresponding colonies but extends to the oppression of all kinds. Remarkably, Said himself expands the scope of "the colonized" to "include women, subjugated and oppressed classes, national minorities, and even marginalized" (*Reflections* 295).

Rob Nixon asserts that the scope of postcolonial studies can be expanded:

> We need scholarship and teaching that can address, in transnational terms, territories beyond postcolonialism's conventional reach. But in so doing we should be watchful that surface geographical gains are not marred by political retreat, that neoliberal acts of violence, for example—especially slow violence—are not hastily euphemized as "global flows." (*Slow* 38)

Though "beyond postcolonialism's conventional reach" which geographically usually covers countries with colonial histories in Africa, Latin America and Asia, China still grapples with issues like environmental justice that is applicable to the postcolonial paradigm. In fact, disadvantaged farmers and ethnic minorities in China, outside the conventional postcolonial reach, are haunted and "Othered" by environmental nationalism ranging from the reluctance to release environmental data to the continual deprival of the powerless' environmental rights. Wu Xiaoqing and other officials deliberately make use of postcolonial discourse to obscure the "unimagined communities" (*Slow* 150). As Vijay Mishra and Bob Hodge argue, postcolonial theory foregrounds "a politics of oppression and struggle, and problematize the key relationships between center and periphery" (276). In this sense, postcolonial theory generally aims to offer resistance to all forms of oppression. Ironically, however, in the context of contemporary China, it has been transformed by the privileged as a strategic discourse and tool to aid in oppression. Historically, in the collective memory of Chinese people, the territorial occupation by Western and Japanese colonizers since the Opium War to WWII is popularly conceived of as an indelible national trauma and humiliation. After the 1989 Tiananmen Square incident, the rhetoric of "western cultural imperialism" or "American hegemony" is evoked strategically by Chinese authorities as a useful discourse to obtain loyalty from the people. Clearly, as Rey Chow maintains, "the People's Republic thus allows itself to hold on to the status of a victim while it openly acts as a militaristic empire. By pointing fingers at 'Western' imperialists, China veils its own *violence* against its people and its colonized territories" (210; italics ours). Such tactics are also currently applied to environmental problems when Wu Xiaoqing tries to deny the Beijing $PM_{2.5}$ report released by the U.S. embassy. In the shadow of transnational environmental justice, the Chinese government's actions not only harm its own citizens' but also potential foreign newcomers' right to know what environment they have lived in or will move into. In the following section, we will see that China also demonstrates such "slow violence" against the subaltern farmers as lower-class and the colonized Tibetans as an ethnic minority, albeit in different tunes.

C. THE BOTTOM CLASS IN ENVIRONMENTAL JUSTICE: TRAPPED IN THE CONTAMINATED MOTHERLAND

In China's developmental path, all the national policies regionally benefit the cities and the industries while rural areas always sacrificially provide natural resources and labor force for industrial capital accumulation in remote cities. The notorious "Great Leap Forward" [*da yuejin*] from 1958 to 1961, led by Mao Zedong and aimed to rapidly transform China from an agrarian economy into a communist society through rapid industrialization and collectivization, is such an example. Mao saw steel production as the key pillars of economic development. He forecasted that within fifteen years of the start of the "Great Leap Forward," China's steel production would surpass that of the UK. To meet this wildly optimistic target, Mao encouraged the establishment of small backyard steel furnaces in every rural commune. Huge efforts on the part of peasants were made to produce steel out of scrap metal. To fuel the furnaces the local environment was denuded of trees and wood taken from the doors and furniture of peasants' houses. Pots, pans, and other metal artifacts were requisitioned to supply the "scrap" for the furnaces. Many of the male agricultural workers were diverted from the harvest to help the iron production. Apart from lumps of pig iron which was of negligible economic worth and disappearance of large areas of forest, the "Great Leap Forward" finally led to the great famine resulting in at least eighteen million deaths, the overwhelming majority being peasants who had no freedom of fleeing from their famine-stricken villages to beg for food in cities.

To make the system of unbalanced growth work, the communist Chinese government actually adopted the toughest residence registration system perhaps in history to prohibit farmers from migrating freely. Even though the first Chinese constitution passed in 1954 has granted that the Chinese can enjoy the liberty and freedom to live and migrate, the implementation of *Regulation on Household* [*hukou*] *Registration of the People's Republic of China* in 1958 cancelled this right by limiting the farmers to the rural areas. Article 16 stipulates:

> Citizens leaving the permanent place of residence for personal reasons and living temporarily elsewhere for longer than three months must apply for extension [of temporary residence] or migration; those with no reason to extend and not qualified to relocate should return to their permanent place of residence.[6]

To meet the demand from cities where the national power and wealth disproportionally centers, environmental degradation originating from mining, deforestation, chemical fertilizers and other human activities in vast rural areas is increasingly acute. When their land is totally contaminated, however, these underprivileged farmers, unlike the environment-worsening metropolis' rich people who can fly to other cities or countries to take a fresh breath, cannot easily escape from their deteriorating environment: they are *unfairly* trapped. Thanks to the discriminatory distinction between rural and urban people, the farmers cannot migrate legally to the much developed urban regions to share social, political and economic goods, resulting in unfair distribution of environmental harm. Chinese writers Yan Lianke and Ma Jian have keenly perceived it in *Streams of Light and Time* and *The Dark Road* respectively.

In *Streams of Light and Time,* Yan Lianke powerfully delineates an epic about the Chinese farmers desperately trapped in Sanxing Village, an isolated area bearing startlingly high levels of fluorine in the water and unknown chemical elements in the soil. Thanks to the toxin, the farmers who are plagued by a laryngospasmic disease cannot live beyond forty years.[7] In order to live longer, four village chiefs strive to get rid of the doom in various ways: the four village leaders, named chronologically, Du Sang, Sima Xiaoxiao, Lan Baisui, and Sima Lan, command their villagers to raise more children, plant only rapeseed on the soil, plow deeply because the deep soil is supposed to be uncontaminated, and irrigate with water from the remote Lingyin canal. Unfortunately, all the efforts are in vain.

Yan does not directly point out the migration spatial restriction imposed upon the farmers under the rigorous censorship practice. Rather, he employs temporal genealogy and historical narratives to foreground the well-known secret. Remarkably, Yan strengthens the image of a destined failure to a clean and safe environment in flashback of four generations, with the characters moving forward to the past. For example, Sima Lan's death in his 40th year appears in the first chapter after the water from the canal turns out to be stinking and dirty. The description about Sima Lan as an embryo in his mother's womb appears in the very end of the book where the trauma of elder generations is delineated as well. In other words, this is a story from death to life. Consequently, the aesthetic repercussions can be perceived. This reversed structure is powerful not only in evoking empathy for the farmers' harsh subsistence and hostile environment, but also in articulating the genesis of the villagers' environmental doom which is intergenerational. In short, the fictionally reversed order seems to forcefully tell us that the environmental injustice has been deeply rooted long since.

Yan coins the term "deity-realism" to account for the doom. According to Yan, "deity-realism" [*shenshi zhuyi*] (*Discovering* 181) forsakes the obvious and direct logical connection in reality to explore the invisible and hidden truth. Yan observes that "logic inside" [*nei yinguo*], transcending the boundaries of all the traditional causes and effects, indeed contributes to the unique charm of "deity-realism." Yan further criticizes that "some writers only sniff the stink by the riverside; some authors only describe the poisonous liquid. With the tongue of literature, blaming men who are not what they were in the times of long ago to simply talk about China and Chinese people cannot fully comprehend the plight of people nowadays" (*Discovering* 183). *Streams of Light and Time* is a book that delves into the absurd existence in a harsh environment with the "logic inside." Yan's "logic inside" is well founded. Historically, the reason why Yan chooses writing as a career is that writing enables him to "leave the land and leave the countryside so that I will no longer toil in the farm any more like my parents and sisters" (*My Reality* 12). Bearing in mind the "deity-realism" and "logic inside," Yan compels the readers to reconsider the fundamental reasons of the entire misfortune: since the implementation of the discriminatory migration policy in 1958, farmers have been restricted to residence in isolated villages even in the twenty-first century. Yan suggests that their lives are actually predetermined by some grand political-economic structures among which the most visible household registration rules and other invisible discriminatory policies have effectively trapped these farmers in their contaminated motherland.

Apart from the inversed chapters, Yan's literary presentation of a clean environment on the locked rural land incorporates a parody of literary classics, which are meticulously organized and situated. In its degree of isolation, Sanxing Village is akin to the Peach-Blossom Spring, a Utopia described by the ancient Chinese poet Tao Yuanming during the Eastern Jin Dynasty (317–420 AD), when China was in turmoil. In the narrative, a fisherman serendipitously wanders into a lesser known area with blossoming peach trees, where villagers tell him that their ancestors escaped to this location in the Chin Dynasty (221–206 BC) to avoid the war and that they have cut themselves off from the outside world since then. In this utopia, animals and human beings live in harmony: "the fowls and dogs of one farm could be heard from the next" (Tao 167). Indeed, Peach-Blossom Spring reflects political evasiveness by an imaginative retreat into an isolated and clean environment which serves as a haven for the dispossessed farmers.

Sanxing Village, though not completely secluded from the outside communist clout, can also be called a modern "Peach-Blossom Spring" in that Lan Baisui, for example, claims that "there is nobody to ask them to join a meeting in the past decades" (*Streams* 228). Dissimilar to the world outside wherein a village leader will be a member of the Communist Party and people are judged by their class origins (*Streams* 260), Sanxing Village has never heard of these before, just like those who live in Peach-Blossom village "had never even heard of the Han Dynasty, let alone its successors the Wei and the Chin Dynasty" (Tao 167–68). However, Sanxing Village is not so much a haven as a forgotten, contaminated corner. Though the historical amnesia is analogous, the startlingly high level of fluorine in water and unknown chemical elements in the soil underscores the tragic fate of the farmers in Sanxing Village.

While the farmers in Peach-Blossom Spring intentionally escape to the outside world, the farmers in Sanxing Village are purposely marginalized and trapped in their native soil by the world outside. Lan Baisui, whose first name means to live for a hundred years, sincerely holds the belief that Director Lu, the Communist leader, and his corresponding subordinates can inspire villagers to plow deeply thereby reducing exposure to contamination and enabling villagers to live longer. Yet, the farmers are in fact are pushed around by Director Lu, who later leads the great escape of all the village officials from Sanxing Village because "Sanxing village is not marked on the map of the county and its governance is not clarified" (*Streams* 304). More important than what Yan writes is what he omits: a Peach-Blossom Spring where the environment is clean and the power is fairly distributed is impossible to realize under the Communist regime which paradoxically advocates another grand utopian project—realizing the Communist goal for all the oppressed.

Yan's literary trope regarding the land bondage is intensified in chapter 4 when Sima Xiaoxiao tries to lead the villagers to survival during the plague of locusts in the 1950s. Yan quotes *Exodus* in the Bible as the epigraph in each section of this chapter. Moses's active struggle towards the Promised Land from the oppressed Egypt contrasts sharply with the despair of villagers trapped in Sanxing Village. During the great famine period, the helpless Sanxing farmers have eaten up even the seeds of the trees. When the seeds are distributed, the disabled and sick children are abandoned deliberately; this painful decision is taken in order to feed the other children and adults. As Sima Xiaoxiao puts forwards this method, even though he is aware that three of his children will be starved to death, he is helpless, unlike Moses who can evoke aid from God:

> "I don't want to die, Uncle Xiaoxiao. I am only seven years old."
> Xiaoxiao Sima takes a glimpse at the child.

"Just go ahead, and it's your filial respect to your parents." (*Streams* 375)

Sima Xiaoxiao is not insensitive enough to evade the consequences of the traumatic decisions. His intentional ignorance belies his sorrow deeply ingrained in his heart when he expounds the concept of filial reason to the lame child. Even though filial piety is highly valued in China, it is still a little ironic that adults who ought to care for their children have to urge their disabled children (a burden in famine) to die as soon as possible so that the extremely limited rationing food can sustain the survival of the adult. The sacrifice of the disabled is later in fact transformed into quasi-cannibalism. The crows eating the carrion of the disabled are eaten by the surviving farmers, resulting in a stark contrast with Tao Yuanming's *Peach Blossom Spring* where a benign relationship between human beings and animals can be perceived. Even though the disabled are abandoned, dispossessed farmers continue to starve to death. Yan reinforces his criticism when he contrasts the abject villagers' starvation due to the famine coupled with the laryngospasmic disease sharply with the epigraph describing the fortune of Moses to obtain the badly needed food from Jehovah. While Moses eventually arrives in the land of milk and honey, Sima Xiaoxiao is doomed to failure as he and other farmers are not entitled to leave the poisoned land. Land epitomizes the contested ground for the institutional discrimination and land is also the location where officials exercise their authority.

While Yan Lianke's *The Streams of Light and Time* elucidates the consequences of the exclusion for those who are trapped, Ma Jian's *The Dark Road* enlightens the outcomes of "discipline and punish" for those who flee. When it comes to the writings of the fugitives, it will make some sense here to bring *The Adventures of Huckleberry Finn* into our sight where the same theme is embodied as well. *The Adventures of Huckleberry Finn* and *The Dark Road* are both about an escape from the dehumanized population control regulation: Black boy Jim escapes from the slavery in the southern states of the United States, while Kongzi and Meili escape from the Chinese family planning and household registration systems; Jim flees the on Mississippi River while Kongzi and Meili go down the Yangtze River. Indeed, Ma suggests in *The Dark Road* via the delineation of a Chinese couple's harsh journey that Chinese family planning policy is in effect modern slavery. *The Dark Road* is also an odyssey by waterway by which pregnant Meili and her husband Kongzi, the 76th-generation descendent of Confucius, flee with their young daughter to escape the atrocious one-child policy. After a painful and long journey, they finally come to a place where the one-child policy is not enforced strictly because the local air is so dirty that it can even kill a man's sperm.

The environmental injustice is mirrored on the bodies "which are not only the sites of the direct application of power, but permeable sites that are forever transformed by the substances and forces" (Alaimo 30). For example, the title of Ma's novel is punny: on the one hand, the Chinese name of *The Dark Road, Yin Zhi Dao,* is similar to yindao, which denotes Meili's reproductive organs in Chinese, where she is not able to assume her valid control as "the state controls our wombs" (29); on the other hand, the dark road is a metaphor to imply that the Kongzi and Meili's escape along the waterway is very tough and stealthy probably without a promising future even when they in the end arrive at the so-called Heaven Township [*tiantang zhen*]. Ma's *The Dark Road* offers potent avenues for the bodies that are entangled with brutal politics, segregation laws and toxic economy. Ma has concretizes this intertwinement vividly during the end of the Kongzi and Meili's journey:

> They [Kongzi's family members] . . . sailed downstream, picking up and delivering cargos as they went, until they reached the dirty industrial town of Dexian in Western Guangdong Province, where they anchored for the last week. Although Kongzi was able to pick up delivery jobs there, it was not a pleasant place to stay. At night the paper factories would spew into the river foul waste water that smelled of rotten shrimp paste and caused the three of them to cough and gag in their sleep. (148)

By noting the relationship between water and land, Ma captures the paradox of the land and water for the farmers here. Kongzi's family, who is supposed to till the fields, has to be displaced for being prosecuted by the one-child policy. Even though Kongzi can leave his boat to work on land later, he still worries about the police on land. Life on water thus becomes an alternative haven for the abject family. However, the water road is not a paradise that can be regained: it is also a dark road. As the novel demonstrates, policemen finally find the pregnant Meili on the boat and force her to undergo a violent abortion. Moreover, life on the boat is haunted by the foul waste water emitted from factories. In other words, life on the water is not free from the control of the land given that the water is land-locked. In this sense, the farmers are "Othered" as homeless creatures in the motherland, regardless of locus. In China, "motherland" [*zuguo muqin*] is repeated over and over again to proclaim the blood ties between the country as the mother and the people as children. Everyone should love the country just like they love their mothers. Patriotism strengthen by the moral discourse of motherland is very useful to evoke unification between different social groups to fight against drought, flood, other internal and external threats, thus disseminating an overpowering atmosphere that anyone cannot dissent from. Yet, in reality, the biological mother is chased by the authoritarian mother that deprives the biological mother of the right to live happily on her own soil. The biological mother and authoritarian one are not unified thanks to toxic chemicals and the brutal migration policy. This division has resulted in a rhetorical irony for the proclaimed slogan. Meili's body, biologically and epistemologically, is a passive contested site where power with regard to the migration restriction, one-child policy and the industrial pollution exert its influences.

Ma confirms the environmental degradation for the "Othered" farmers: "Although all the workers are illegal migrant peasants, no government official would be willing to brave the stench to come and check their documents" (219). Ironically, it is the badly polluted environment that offers the unprivileged an opportunity to partly escape the control of Big Mother. In this immediate context, the "motherland" is transformed into a landfill site which is no more than a deserted corner, like Yan's Sanxing Village. The poor are protected not because they are highly valued but instead because they are shielded by the toxic industrial waste that is irresponsibly abandoned in their habitats, implying that the dispossessed poor are just like the squalid waste.

While Ma in *The Dark Road* resorts more to ethos to shed light upon the marginalized farmers, Yan employs pathos to discuss the interplay of gender and environmental justice in *Streams of Light and Time*. Sima Lan asks Lan Sishi, his lover, to engage in the "flesh trade" to raise money to build the canal so that the Sanxing Village can get clean water from sixty kilometers away. Lan Sishi finally comes into the town to be a prostitute. This scenario is strikingly tragic:

> The room is dark, like the rainy days. . . . She has been offering sexual services to men, unremittingly, ever since she arrived in the town from the Palou Mountain. Finally, after the man scratches her hair at will to fulfill his pleasure, he puts on his trousers, leaving a hundred RMB behind. (*Streams* 50)

Yan makes a parallel between the spatial distance from the rural mountain to the urban town with the temporal distance for Lan Sishi's sex work. For Lan Sishi, the only way to live up to expectations is to obtain adequate money to raise fund for Sima Lan, who leads the farmers to construct the "Lingyin" canal, which literally means the hidden soul in Chinese. In fact, the hidden soul is absent; eighteen farmers die and twenty-one are handicapped, thanks to the project. To the farmers, the completion of the canal means "Mingtong" (*Streams* 50), which literally means that the good fate will come through. In this sense, Lan's sex work pursues the noble goal of offering a clean environment for future generations.

Yan articulates the marriage between gender, revolution and environmental justice in rural areas. Lan Sishi has to lose her virginity to Director Lu for the help he renders to the Sanxing Village. Apart from women's vaginas, men's skin is also a site where the "revolutionary power" acts. The needs of the revolution were expediently used to justify all sorts of sacrifices made by ordinary people. During the Cultural Revolution, men had to flay and sell part of their skin in the hospital to make money, but their financial rewards were denied in the name of revolutionary slogans. The slogans include "serve the people" [*wei renmin fuwu*] (*Streams* 247) and "when one joins the revolution, it is very common to bleed" [*geming zonghuiyou liuxue, zonghuiyou xisheng*] (*Streams* 248). Remarkably, the revolutionary discourse is treated as an excuse to deny farmers' efforts to change their toxic environment via the direct exploitation of their skins.

At the end of this section, it should be noted that Ma's scope is wider than Yan's. In Ma's *The Dark Road*, Kongzi's final destination is polluted by the toxic chemical waste imported from western countries which has resulted in malformation in neonates and low reproduction rates. The practice of Western developed countries shipping their waste to developing and underdeveloped nations for disposal produces environmental injustice on an international scale, which is a classic North-South debate. The transnational environmental inequities mediated by the Chinese undemocratic system tend to exacerbate the environmental risks confronting ordinary people in China, especially the poor and other politically powerless groups. In this way, Ma connects transnational and domestic environmental injustice. Furthermore, Ma covers more diversified issues such as educational inequality and food security. That Kongzi's daughter cannot receive education because she comes from a rural region is a striking irony given that Kongzi's ancestor, Confucius, is hailed as the greatest sage and teacher in Chinese history. Besides, the poisonous powdered milk that renders children's disabilities in the novel also touches upon the hot topic of food security in today's China. Ma echoes Graham Huggan and Helen Tiffin's argument that there is "no social justice without environmental justice; and without social justice—for all ecological beings—no justice at all" ("Green Postcolonialism" 10) and meanwhile demonstrates the "integrated approach" (19) to environmental justice proposed by Adrian Armstrong.

D. ETHNICITY IN ENVIRONMENTAL JUSTICE: TIBET AS AN OCCUPIED LAND

Unlike Yan and Ma who writes about the "Othered" farmers who are not allowed to migrate to other livable environments, A Lai portrays the Tibetans who themselves don't want to migrate but are invaded on their territories.[8] The immigration of Han Chinese who makes up 92 percent of China's population has brought about a new lifestyle that conflicts with the Tibetan traditional ethnic understandings and practices stemming from their local ecology. To some degree, it results in a so-called "ecological imperialism" or "green imperialism."[9]

In "Heaven Fire," which is one section of *Empty Mountain*, Duoji is an ancestral wizard in the village, a Tibetan religious figure. In the eyes of

Cunji villagers, Duoji can predict the direction of the wind and set fire to eradicate the rotten material without burning the forest. Every year, in order to make the green grass grow better to feed the cattle, Duoji leads other farmers to set the fire. The Han policeman, regardless of the Tibetan tradition, arrests Duoji, but he is bailed out by the Tibetan villagers. One early winter day, Duoji is caught again because of the fire. He finds that the street in the town is teeming with red banners like the wildfires, and has lost its control in a chaotic and violent manner. It turns out that the Cultural Revolution has broken out. Duoji is treated as an anti-revolutionary enemy this time. Policeman Wei is removed from office as he is accused of showing sympathy to Duoji. After the political campaign against Duoji, he escapes from the prison van, hiding in a cave. At the same time, a fire without definite causes spreads across the forest, approaching Cunji Village. During the firefighting, conflicts break out among the Tibetan Cunji Village and the Han Chinese firefighters. Han engineers release the lake water to put out the fire but neglect the religious connotations of the water. The lake disappears, and the fire burns more savagely.

The paradox of the land is also revealed by A Lai. The Han Chinese regard the land as a national property, where the woods and the minerals can be exploited at will. "The Tibetan villagers don't understand why the forests have an owner named country in a single night, even though Tibetan ancestors have lived here for hundreds of years" (*Empty* 142). The Han Chinese holds that all the resources are possessed by the country—in fact, by the Han Chinese. Paradoxically, after the fire breaks out, the Han Chinese conceive of the woods as the property of the Tibetans when the Han Chinese want to get food from the Tibetans. To the Han Chinese, they just come to put out the fire to help the Tibetans. A Lai wants to show that the coming of the Han Chinese means depletion of natural resources. In fact, the firefighters become woodcutters after the fire has ended. Local Tibetans cannot do anything to stop the resources from being utilized, and they do not stand to gain from the exploitation. Such environmental damage is extremely tragic not only for the local natural ecology but for the soul and spirit of the Tibetans because the Tibetan society has focused on sustainability, with its ethnic culture and religious beliefs forbidding the unnecessary killing of animals and excessive logging (Lowe 1992). A Lai's narration in some sense echoes the Nigerian Ogoni people's protest against "ecological genocide" due to the invasion of western oil companies.[10] Since the late 1960s, the mineral deposits in Tibet have been extracted under poor environmental regulations. The exploited minerals, including copper, gold, lead, iron and zinc are of greatest interest to the Han Chinese. The trees are cut down tremendously and hydroelectric power is developed to supply electric power for East China with little benefit for Tibetans. In fact, the Han Chinese's environmental assaults on Tibetans perpetuate a "resource curse" that changes the blessing into misfortune.[11]

A Lai addresses the ecoambiguity of Lao Wei, a Han policeman who displays contradictory attitude toward Duoji who has a deeper understanding of the indigenous environment. On the one hand, he open-mindedly sets Duoji free when Duoji promises not to set fire again. He alerts other Han Chinese to the pressing needs of extinguishing the fire to "save the trees" (*Empty* 235) rather than hold the meaningless meetings. As such, he shows his willingness to protect Duoji who understands nature well. On the other hand, as a Han policeman, he seduces the Tibetans to expose the location of the exiled Duoji, trying to arrest Duoji even when he himself is without official duties. In this sense, Lao Wei paradoxically betrays his reluctance to support Duoji. At root, such ecoambiguity results from the collision between Tibetan ethnicity and socialist politics following the Han's colonialization of Tibet.[12] Just as Raymond Williams contends that "the idea of nature contains, though often unnoticed, an extraordinary amount of human history" (*Problems* 67), the divergent perceptions on the eco-environment between Han Chinese and Tibetans can be examined in their separate human history: "For the Han Chinese, taking up the land and conquering nature become central goals: There is endless joy in fighting with heaven; there is endless joy in fighting with the earth; there is endless joy in fighting with the human" (*Empty* 194). It is a practice advocated by Chairman Mao: nature is subject to human control especially during the Cultural Revolution. In Mao's era, traditional Tibetan culture was "denigrated as 'backward' and 'primitive' in the Marxist-inspired social theory" ("Fragments" 59).When the Han Chinese took power in Tibet in 1959, the Tibetans became the target of liberation by the Han Chinese from backwardness, ignorance and bondage with the indigenous Tibetan monks displaced and the religious leaders prosecuted. In fact, the ethnic "Other" and the environmental "Other" are the same in essence. Before the Han Chinese invaded Tibet, Tibet was not only home to indigenous Tibetans who already knew the terrain intimately and had their own mythologies rooted in their dwelling places but also home to a rich variety of wild animals and plants. However, when the Han Chinese arrived, the Han Chinese showed a parallel colonization of people and physical environment. In other words, the Tibetan religion and language, together with the wilderness, is "Otherized" by the Han Chinese.

The Tibetan religion is a vital factor highlighted by A Lai because it is closely linked with Tibetan environmental notions, and the elimination of the religious practices by Communist China has resulted in great harm to the environment. Bon, a native religion in Tibet, regards the water, woods, animals and mountains as holy. In Bon doctrine, every practice should not be inconsistent with the god.[13] Thus, this worldview has made a great difference in reality. The bear, for example, which is the traditional target of the hunter, doesn't kill the hunter even though it is able to do so (*Empty* 203); the hunter doesn't shoot the bear in return. A Lai suggests that the hatred between the bear and the hunter has almost become one of care and concern due to the fact that the Tibetans perceive that the animals can communicate with human beings. A traditional relationship between predator and prey has been subverted through religious explanations. "Gold Wild Duck" [*jin yeya*] is hailed as the god who protects the Cunji village and the forest functions as the communication channel between the earth and heaven. Besides, fire and water are regarded as gods by the Tibetans. Tibetan ancestors believe that water is pure and is the essential resource for the survival of the world at large. The water god is in charge of floods; it can also bring about disaster to the prairie lands, gorges and rivers, overwhelming everything. Thus, the Tibetans conceive of water as holy. The fire god that offers light and heat is also carefully worshipped so that it will not cause devastating fires. The Tibetans understand deeply why the fire can pass through the river as they know that the original balance between the fire god and water god is disturbed. Worster contends that a river can reveal much on the "flow of power in history" (*Rivers of Empire* 17). The original Tibetan lifestyle, including the religion, is displaced by the Han Chinese who exert their power on the Tibetan landscape. A Lai employs "heaven fire" [*tianhuo*], a fire from heaven, to denote the fire without human causes to evade the Chinese governmental censorship. In the 1960s, when the monks are prosecuted, and the temples are closed, the concepts of natural gods are denounced as "feudal misbelief" [*fengjian mixin*] (*Empty* 260) by Han Chinese. In this sense, the coming of the Han Chinese, who eradicate the traditional religion that respects nature as holy, is responsible for that fire from heaven. Homi K. Bhabha argues, "The objective of colonial discourse is to construe the colonized as a population of degenerate types on the basis of racial origin, in order to justify conquest and to establish systems of administration and instruction" (70). The construction of the degenerate types for Tibet is in part fulfilled in religious and linguistic ways. While the religious persecution on the Tibetan land is visible, the linguistic transformation is perhaps imperceptible. A Lai shows the linguistic displacement of Tibetan language bearing the notion of endemic environment, and the consequences of such alienation upon the native Tibetans:

> The Blue Overalls smiled: "Do you call the sun as wild duck?"
> ". . . Your Mandarin is bad, I do not understand your language; so I want to ask you, do you call the sun's reflection as the wild duck?"
> Yang Jin shook his head.
> "Yeah, even though your language is dull, you would not come to the extent of joining those irrelevant together . . ." (*Empty* 252)

The Blue Overalls, a Han Chinese, casts his doubt on the Tibetan native language. The Blue Overalls represents the Han workers who come to Tibet. What these workers bring about is not only the industrial lifestyle, but a new language. The Blue Overalls suspects the meaning of the "Gold Wild Duck." He uses the "wild duck" to denote the "Gold Wild Duck," showing his ignorance of Tibetan religious tradition. More importantly, he presupposes that Mandarin, the official Han language, should be the one with which to communicate with the Tibetans. Yang Jin, a Tibetan, does not respond to him when he speaks Mandarin. The Blue Overalls' ignorance has further developed into contempt because of his accusation of Tibetan's language as "dull." As such, a linguistic hierarchy is established as Tibetans cannot defend their indigenous language. Such a linguistic hierarchy actually contributes to a hierarchy among humans. Kincaid argues that "this naming of things is so crucial to possession—a spiritual padlock with the key thrown irretrievably away—that it is a murder, an erasing" (122). As such, the renaming of the golden duck, under the secular Han's domain, is a practice of discipline and control. Foucault uses the term "power/knowledge" to signify that power is constituted through accepted forms of knowledge, scientific understanding and "truth."[14] To the Han Chinese, the "truth" is to follow Chairman Mao's call to conquer nature rather than to live in accordance with it. Therefore indigenous ethnic understanding and knowledge systems are totally overlooked. Thus, the Mandarin language justifies the Han Chinese people's power to transform the environment at their will. A Lai clarifies the difference:

> Mandarin represents newspapers, textbooks, televisions, townships, officials, science and other novel but powerful entities. The Tibetan language is inborn, encompassing religion, nomadic life patterns, farming, household stories, memory and emotion. It was the language landscape itself, forming the contrast of primitive and modern, the official and the folk, science and superstition, progress and stagnation. ("Hanyu" 18)

The ethnic minority, under the control of the Han Chinese, cannot preserve its language. A Lai, himself a Tibetan writer, articulates his worry when he sees that the newspaper and the primary school in Tibet prefer Mandarin to Tibetan. Like Stephen Greenblatt's definition of culture, colonization itself employs "an ensemble of beliefs and practices" that "function as a pervasive technology of control, a set of limits . . . a repertoire of models" through which the colonizers further their ultimate goal of possession (225). After Mandarin is adopted by the Tibetans, it seems inevitable that the young Tibetans will gradually forget the "Gold Wild Duck," the water god and fire god, and the human tie to the animals and the whole of nature.

Arif Dirick has suggested that "orientalism . . . in some basic ways required the participation of 'orientals' for its legitimation" (112). In *Empty Mountain*, the participation of the "orientals" is contributed by characters like Suobo, a Tibetan who follows the Han Communist Chinese and denounces other Tibetans that don't follow the Communist rule. When other Tibetans refuse to follow his instruction, he threatens them by saying that "you all are out of date; you all are decadent. . . . You will make big mistake!" (*Empty* 184). Suobo, like other Han Chinese, overlooks the agentic power of fire and reduces it to an arena where he can show his power: he slaps his father who disobeys his order to prove that he controls the village. As a native inhabitant, he acculturates the intrusive Han Chinese notions of progress. In other words, unlike Duoji and Yangjin who passively resist the assimilation, Suobo actively participates in it by abandoning Tibetan ethnic perceptions and practices, embracing everything from the Han Chinese pragmatically as a source of power, and remolding himself consciously. As a result, the legitimization of land occupation, together with environmental notion changes, is in the end fulfilled by the "self-orientalization" of native Tibetans.

E. CONCLUSION

This article seeks to propose and develop the three major dimensions of environmental justice in contemporary Chinese literature. In terms of the transnational dimension, apart from the classic North-South debate, Beijing's $PM_{2.5}$ event reveals an updated fact that new issues such as environmental sovereignty in an information age are complicating matters pertaining to environmental justice especially in those "Orientalized" countries, which sometimes makes the relevant discourses temporally and spatially distorted. As we see, Communist China painstakingly resorts to misplaced postcolonial discourse to withstand increasing pressure from transnational environmental justice issues related to the domestic/global citizens' positive rights for transparent environmental information. Domestically, two dimensions unfold. One is class and the other is ethnicity, both of which are regionalized as rural and ethnic minority administrative units by the political-economic structures of contemporary China. In vast rural areas, farmers are an underprivileged class, both historically and currently, and they bear most of the environmental costs for the industrial development in big cities. Ironically, they cannot easily share the benefits due to a series of discriminatory policies that have lastingly trapped them in their contaminated land. The literary works of Yan Lianke and Ma Jian foreground farmers as China's unique environmental refugees. Yan's discourse is oblique partly due to his conception of "deity-realism" while Ma is more audacious. In the last dimension of ethnicity, the tragic story of Tibet narrated by A Lai illustrates how Tibetan ethnic perceptions and practices related to the environment are suppressed by overwhelming communist and Han Chinese powers. Through the three dimensions, we can see that the concerns about the oppressed, the underprivileged and the disenfranchised stay at the core of the environmental justice ecocriticism. The "oppositional criticism" originating from Said's milestone research is a useful framework to develop new frontiers for ecocritical studies.

Environmental literature is an extremely fertile field because the impact of environmental rhetoric is far from metaphorical. Lawrence Buell observes:

> For technological breakthroughs, legislative reforms, and paper covenants about environmental welfare to take effect, or even to be generated in the first place, requires a climate of transformed environmental values, perception, and will. To that end, the power of story, image, and artistic performance and the resources of aesthetics, ethics, and cultural theory are crucial. (*The Future* vi)

Scott Slovic maintains that "one of the crucial roles of ecocritical scholarship is to help make [environmental writers'] tales audible in the halls of power" (142). We believe the Sinophone literary works analyzed in this paper will call for readers' awareness of environmental justice and its complexities and multiple dimensions, thus leading to possible favorable shifts in the policy-making process in China.

As an attempt to guide second-wave Chinese ecocritical study, this paper sheds light on the oppression and "slow violence" experienced by the poor in China and illustrates the environmental justice motif in contemporary Chinese literature. As Patrick D. Murphy indicates:

> A [transnational ecocritical] theory would transect, that is, cut across, the limitations of national perspectives and boundaries. . . . Avoiding parochialism does not mean practicing universalism, but it also does not mean abandoning the idea that ecocriticism in whatever varied forms it may take is a crucial, relevant, and necessary literary and cultural practice to be promoted worldwide. (63)

The implications of this paper are definitely not restricted to the Chinese context. After all, institutional discrimination and intentional environmental negligence in other places on this planet have much in common. By triangulating three such dimensions of environmental justice as transnationalism, class and ethnicity in contemporary Chinese literature, we hope to offer a new avenue where East Asian studies, ecocriticism, and comparative literature intersect.

ACKNOWLEDGMENTS

We want to thank Rob Nixon, and Karen Thornber for their early advice. We are deeply indebted to Kyhl Lyndgaard and three editors of this collection for their valuable revision suggestions on several drafts.

WORKS CITED

A, Lai. *Empty Mountain*. Beijing: Renmin Wenxue P, 2005. Print.

———."Hanyu: Duoyuan wenhua gongjian de gonggong yuyan (Mandarin Chinese: Public Language Built by Multiple Culture)." *Dangdai Wentan* (*Modern Literary Magzine*) 25.1 (2006):18–20. Print.

Adamson, Joni, and Kimberly N. Ruffin, eds. *American Studies, Ecocriticism, and Citizenship: Thinking and Acting in the Local and Global Commons*. New York: Routledge, 2012. Print.

Adamson, Joni, Mei Mei Evans, and Rachel Stein, eds. *The Environmental Justice Reader: Politics, Poetics and Pedagogy*. Tucson: U of Arizona P, 2002. Print.

Alaimo, Stacy. *Bodily Natures: Science, Environment, and the Material Self*. Bloomington: Indiana UP, 2010. Print.

Armstrong, Adrian. *Ethics and Justice for the Environment*. New York: Routledge, 2012. Print.

Bhabha, Homi K. *The Location of Culture*. London: Routledge, 1994. Print.

Bonfoundation. *Bonfoundation*. Web. 12 Sept. 2013.

Buell, Lawrence. *The Future of Environmental Criticism: Environmental Crisis and Literary Imagination*. Malden, MA: Blackwell, 2005. Print.

———. *Writing for an Endangered World: Literature, Culture and Environment in the U.S. and Beyond*. Cambridge, MA: Harvard UP, 1995. Print.

Cheng, Hong. *Xungui Huangye* (*Return to the Wilderness*). Beijing: Sanlian Shudian, 2001. Print.

Chow, Rey. "Things, Common/Places, Passages of the Port City: On Hong Kong and Hong Kong Author Leung Ping-kwan." *Sinophone Studies: A Critical Reader*. Ed. Shu-mei Shih, Chien-hsin Tsai, and Brian Bernards. New York: Columbia UP, 2013. 207–26. Print.

Clark, Timothy. *The Cambridge Introduction to Literature and the Environment*. New York: Cambridge UP, 2011. Print.

Cohen, Aaron J., et al. "The global burden of disease due to outdoor air pollution." *Journal of Toxicology and Environmental Health, Part A* 68.13-14 (2005): 1–7. 20 May 2005. Web. 13 Sept. 2013.

Crosby, Alfred W. *Ecological Imperialism: The Biological Expansion of Europe, 900-1900*, 2nd ed. New York: Cambridge UP, 2004. Print.

DeLoughrey, Elizabeth, and George B. Handley, eds. *Postcolonial Ecologies: Literatures of the Environment*. New York: Oxford UP, 2011. Print.

Dirlik, Arif. "Chinese History and the Question of Orientalism." *History and Theory* 35.4 (1996): 96–118. *JSTOR*. Web. 7 Sept. 2013.

Dirlik, Arif, and Zhang Xudong, eds. *Postmodernism and China*. Durham, NC: Duke UP, 2000. Print.

Economy, Elizabeth. *The River Runs Black: The Environmental Challenge to China's Future*. Ithaca, New York: Cornell UP, 2010. Print.

Garrard, Greg. *Ecocriticism*. New York: Routledge, 2004. Print.

Greenblatt, Stephen. "Culture." *Critical Terms for Literary Study*. Ed. Frank Lentricchia and Thomas McLaughlin. 2nd ed. Chicago: U of Chicago P, 1995. 225–32. Print.

Grove, Richard H. *Green Imperialism: Colonial Expansion, Tropical Island Edens and the Origins of Environmentalism, 1600–1860*. Cambridge, UK: Cambridge UP, 1995. Print.

Heise, Ursula K. *Sense of Place and Sense of Planet: The Environmental Imagination of the Global*. New York: Oxford UP, 2008. Print.
Huggan, Graham, and Helen Tiffin. "Green Postcolonialism." *Interventions* 9.1 (2007): 1–11. Print.
———. *Postcolonial Ecocriticism: Literature, Animals, Environment*. New York: Routledge, 2010. Print.
Jiang, Steven and Alexis Lai. "China: Haze isn't foreign embassies' business." *CNN.com*. Cable News Network, 6 June 2012. Web. 12 Sept. 2013.
Kincaid, Jamaica. *My Garden*. New York: Farrar, Straus and Giroux, 1998. Print.
Lu, Shuyuan. *Shengtai Piping De Kongjian* (*Space for Ecocriticism*). Shanghai: Huadong Shifan UP, 2006. Print.
Ma, Jian. *The Dark Road*. Trans. Flora Drew. New York: Penguin, 2013. Print.
Foucault, Michel. *Discipline and Punish: The Birth of the Prison*. New York: Vintage, 1994. Print.
———. *The Archaeology of Knowledge*. New York: Routledge, 2002. Print.
———. *The Order of Things: An Archaeology of the Human Sciences*. New York: Vintage, 1994. Print.
Mishra, Vijay, and Bob Hodge. "What is Post(-)colonialism?" *Colonial Discourse and Post-Colonial Theory*. Ed. Patrick Williams and Laura Chrisman. New York: Columbia UP, 1994. 276–79. Print.
Moser, Adam. "The Yangzonghai Case: Struggling for Environmental Justice." *China and the Environment: The Green Revolution*. Ed. Sam Geall. London: Zed Books, 2013. Print.
Murphy, Patrick D. *Ecocritical Explorations in Literary and Cultural Studies: Fences, Boundaries, and Fields*. Lanham, MD: Lexington Books, 2009. Print.
Nixon, Rob. "Environmentalism and Postcolonialism." *Postcolonial Studies and Beyond*. Ed. Frederick Cooper, Laura Chrisman, Ania Loomba, and Jed Esty. Durham, NC: Duke UP, 2005. 233–51. Print.
———. *Slow Violence and the Environmentalism of the Poor*. Cambridge, MA: Harvard UP, 2011. Print.
Pedersen, Ole W. "Transatlantic Movements of Justice: A Story of Inspiration and Diversity." *Environmental Justice* 2.1 (2009): 35–41. Print.
Rojas, Carlos. "Danger in the Voice: Alai and the Sinophone." *Sinophone Studies: A Critical Reader*. Ed. Shu-mei Shih, Chien-hsin Tsai, and Brian Bernards. New York: Columbia UP, 2013. 296–303. Print.
Said, Edward. *Culture and Imperialism*. New York: Vintage, 1993. Print.
———. *Orientalism*. New York: Vintage, 1979. Print.
———. *Peace and Its Discontents: Essays on Palestinian in the Middle East Peace Process*. New York: Vintage, 1996. Print.
———. *Reflections on Exile: and Other Literary and Cultural Essays*. 1984. Cambridge, MA: Granta Books. 2001. Print.
———. *The World, the Critic and the Text*. Cambridge, MA: Harvard UP. 1983. Print.
Saro-Wiwa, Ken. *Genocide in Nigeria: the Ogoni Tragedy*. Reissue. London: Saros International Publishers, 2005. Print.
Schiaffini, Patricia. "On the Margins of Tibetanness: Three Decades of Modern Sinophone Tibetan Literature." *Sinophone Studies: A Critical Reader*. Ed. Shu-mei Shih, Chien-hsin Tsai, and Brian Bernards. New York: Columbia UP, 2013. 281–95. Print.
Schlosberg, David. *Defining Environmental Justice*. Oxford, UK: Oxford UP, 2007. Print.
Shapiro, Judith. *China's Environmental Challenges*. Malden, MA: Polity Press, 2012. Print.
———. *Mao's War against Nature: Politics and the Environment in Revolutionary China*. Cambridge, UK: Cambridge UP, 2001. Print.
Sheng, Anfeng. "Traveling Theory, or, Transforming Theory: Metamorphosis of Postcolonialism in China." *Neohelicon* 34.2 (2007): 115–36. Print.

Shih, Shu-mei. "Introduction: What is Sinophone Studies?" *Sinophone Studies: A Critical Reader*. Ed. Shu-mei Shih, Chien-hsin Tsai, and Brian Bernards. New York: Columbia UP, 2013. 1–16. Print.

Slovic, Scott. *Going Away to Think: Engagement, Retreat, and Ecocritical Responsibility*. Reno: U of Nevada P, 2008. Print.

Spivak, Gayatri C. *The Spivak Reader: Selected Works of Gayatri Chakravorty Spivak*. Ed. Donna Landry and Gerald Maclean. New York: Routledge, 1996. Print.

Tao, Yuanming. "Peach Blossom Spring" *Anthology of Chinese Literature, Volume I: From Early Times to the Fourteenth Century*. Ed. Cyril Birch. New York: Grove Press, 1965. 167–68. Print.

Thornber, Karen L. *Ecoambiguity: Environmental Crises and East Asian Literatures*. Ann Arbor: U of Michigan P, 2012. Print.

United Nations. Vienna Convention on Consular Relations. *Treaty Series*. 1963. Web. 12 Sept. 2013.

———. Vienna Convention on Diplomatic Relations. *Treaty Series*. 1961. Web. 12 Sept. 2013.

Varsava, Jerry. "Jiang Rong's *Wolf Totem*: Toward a Narrative Ecology of the Grassland of Contemporary Inner Mongolia." *Interdisciplinary Studies in Literature and Environment* 18.2 (2011): 283–301. Print.

Wang, Fei-Ling. *Organizing through Division and Exclusion: China's Hukou System*. Stanford, CA: Stanford UP, 2005. Print.

Wang, Ning. "*Chinese Literary and Cultural Trends in a Postrevolutionary Era.*" *Comparative Literature Studies* 49.4 (2012): 505–20. Print.

Wang, Nuo. Shengtai Piping Yu Shengtai Sixiang (*Ecocriticism and Ecologism*). Beijing: Renmin P, 2013. Print.

Wang, Ruifeng. "Cancer Self-help Water Pollution Villagers Killed Two Thousand People a Year."*Xinjingbao*. Web. 23 July 2013.

Wang, Shudong. *Shengtai Yishi Yu Zhongguo Dangdai Wenxue* (*Ecological Consciousness and Contempoary Chinese Literature*). Beijing: Zhongguo Shehui Kexue P, 2008. Print.

Wang, Xirong, et al. *Shengtai Piping Shiyuxia De Zhong Xiandangdai Wenxue* (*An Ecocritical Perspective into Modern and Contemporary Chinese Literature*). Beijing: Zhongguo Shehui Kexue P, 2009. Print.

Wei, Qingqi. *Luxiuzi Wuqilai: Dui Shengtai Piping De Chanfa Yanjiu* (*Green-clad Scholarship: an Interpretive Study of Ecocriticism*). Nanjing: Nanjing Shifan UP, 2010. Print.

Williams, Raymond. *Problems in Materialism and Culture: Selected Essays*. New York: Verso, 1980. Print.

Worster, Donald. *Rivers of Empire: Water, Aridity, and the Growth of the American West*. New York: Oxford UP, 1985. Print.

Wu, Wencong. "Embassies 'break rules' on Air-quality Measurements." *China Daily*. 6 June 2012. Web. 23 July 2013.

Yan, Lianke. *Faxian Xiaoshuo* (*Discovering Fiction*). Tianjin: Nankai UP, 2011. Print.

———. *Riguang Liunian* (*Streams of Light and Time*). Beijing: Beijing Shiyue Wenyi P, 2009 [Huacheng P, 1998]. Print.

———. *Wode Xianshi Wode Zhuyi* (*My Reality and My Realism*). Beijing: Renmin UP, 2011. Print.

Yang, Jincai. "Environmental Dimensions in Contemporary Chinese Literature and Criticism." *East Asian Ecocriticisms: A Critical Reader*. Ed. Simon C. Estok and Won-Chung Kim. New York: Palgrave, 2013. 187–204. Print.

Yue, Gang. "Fragments of Shangri-La: 'Eco-Tibet' and Its Global Circuits." *Postcolonial Green: Environmental Politics and World Narratives*. Ed. Bonnie Roos and Alex Hunt. Charlottesville: U of Virginia P, 2010. 49–63. Print.

———. Rev. of *Red Poppies: A Novel of Tibet*, by A Lai. Trans. Howard Goldblatt and Sylvia Li-chun Lin. *Modern Chinese Literature and Culture*. Apr. 2005. Web. 26 Aug. 2013.

NOTES

1. Air pollution is associated with a broad spectrum of acute and chronic health effects. As part of the World Health Organization (WHO) Global Burden of Disease Comparative Risk Assessment, Cohen Aaron J., et al. (2005) investigated the burden of disease attributable to urban ambient air pollution and estimated that "ambient air pollution, in terms of fine particulate air pollution (PM2.5), causes about 3 percent of mortality from cardiopulmonary disease, about 5 percent of mortality from cancer of the trachea, bronchus, and lung, and about 1 percent of mortality from acute respiratory infections in children under 5 yr, worldwide. . . . This burden occurs predominantly in developing countries; 65 percent in Asia alone."

2. For details of cancer village and pollution in Huai River in China, see Elizabeth Economy's *The River Runs Black*, 71–73 and 88. Visit the website to see the news report on Wu Xiaoqing, available at http://www.chinadaily.com.cn/cndy/2012-06/06/content_15477009.htm and news report on Shenqiu in English http://www.best-news.us/news-4955527-Cancer-self-help-water-pollution-villagers-killed-two-thousand-people-a-year.html whose original Chinese edition is available at http://www.bjnews.com.cn/opinion/2013/07/20/274242.html.

3. See Cheng Hong's *Xungui Huangye* (*Return to the Wilderness*, 2001), Wei Qingqi's *Luxiuzi Wuqilai: Dui Shengtai Piping De Chanfa Yanjiu* (*Green-clad Scholarship: an Interpretive Study of Ecocriticism*, 2010), and Lu Shuyuan's *Shengtai Piping De Kongjian* (*A Space for Ecocriticism*, 2006).

4. See *The Spivak Reader—Selected Works of Gayatri Chakravorty Spivak*. Spivak argues that "strategic use of positivist essentialism in a scrupulously visible political interest" (215). She argues that it is sometimes advantageous for people to temporarily "essentialize" themselves, making a simplified way to achieve certain goals though there might be differences among the groups. It is an approach of using group identity as a basis of struggle while also debating issues related to group identity within the group.

5. To have an overview of postcolonial studies in China, see Sheng Anfeng's *"Traveling Theory, or, Transforming Theory: Metamorphosis of Postcolonialism in China"* (115–36). He contends that "I see no reason that the intellectuals should give up their right to criticize the country if the criticism is honest and constructive. The concern for the nation and the country is not the infinite extolment of nationalism. Love is sometimes embodied in fierce criticism" (127). His colleague, Wang Ning, entertains the idea that Chinese scholars should not be isolated from the western counterpart. See Wang Ning's *"Chinese Literary and Cultural Trends in a Postrevolutionary Era."* See *Postmodernism and China* edited by Arif Dirlik and Zhang Xudong for further reading on the Chinese reception of western critical theories.

6. See Wang Fei-Ling's *Organizing through Division and Exclusion: China's Hukou System* (87) for a detailed analysis of Chinese household registration system.

7. In fact, the laryngospasmic disease is probably cancer due to high fluorine in water and a possible unknown radioactive chemical element whose atomic number is after 126.

8. To put it precisely, it refers to the Greater Tibet (or Cultural Tibet), including Tibet Autonomous Region, Qinghai Province, Yunnan Province, and Sichuan Province. *Empty Mountain* takes place in Sichuan. A Lai was also born in Sichuan Province where Han Chinese and Tibetans coexist. He is born of Tibetan and Hui parents. The Muslim Hui minority is another of the fifty-six recognized minorities in China. Literary critiques call him "'hybrid' twice over." See Yue Gang, http://mclc.osu.edu/rc/pubs/reviews/yue.htm, and Carlos Rojas' "Danger in the Voice: Alai and the Sinophone". For a comprehensive understanding of Tibetan literature, see Patricia Schiaffini's "On the Margins of Tibetanness: Three Decades of Modern Sinophone Tibetan Literature."

9. Alfred Crosby coins "ecological imperialism" to illuminate the biotic entity that allows the expansion of Europe. See *Ecological Imperialism*. Here, we mean the Han

Chinese have upset the established Tibetan environment. Similarly, Richard Grove names "green imperialism" to offer the etymological definition of diaspora as the spread of seeds and disassociates flora and fauna with a natural landscape.

10. "Ecological genocide" is linked with Ken Saro-Wiwa and his Nigerian colleagues who are against the environmental degradation and oil development resulting from Federal Republic of Nigeria and the oil companies in Niger Delta. See Rob Nixon, "Environmentalism and Postcolonialism" (233–34), and Saro-wiwa, *Genocide in Nigeria*.

11. See Rob Nixon's *Slow Violence and the Environmentalism of the Poor*, especially chapter two, for an insightful analysis of the resource curse.

12. See Shu-mei Shih's "Introduction: What is Sinophone Studies?" in which she argues that Chinese colonization, unlike Europeans who take the oceanic paths to expand, adopts the "continental colonialism" which means that the territory is expanded via continent. For instance, China enlarges its territory in Xinjiang, Mongolia and Tibet through the continent in Qing Empire during the eighteenth century.

13. There are disputes on the religious affiliation of Bon. Tenzin Gyatso, the 14th Dalai Lama, acknowledges Bon as the native religion of Tibet. The ancient root of the Bon religion stems from a profound respect for nature and highlights the healing of physical and environmental as well as spiritual afflictions. As Indian Buddhism was being established in Tibet, many native Bon elements were incorporated into the incoming religion, resulting in a distinct religion known today as Tibetan Buddhism. In turn, Buddhist influences are abundantly evident in the Bon religion as it currently exists. The two religions are distinct in many ways but share a strong and identical commitment to bringing an end to all suffering. See http://www.bonfoundation.org/aboutbon.html for Bon's history and religion.

14. See Michel Foucault's *Discipline and Punish*, *The Archaeology of Knowledge*, and *Order of Things* for detailed discussion of the interplay of power and discourse, discipline and punishment.

FOUR

Minamata and the Symbolic Discourse of the South

Tsutomu Takahashi

Whether Japan and its geographical and economic factors should be considered under the rubric of the Global South may be controversial. Geographically, the Japanese archipelago is situated on the periphery of East Asia at the northern latitude of 30–45 degrees; its natural climate enjoys four distinct, variegated seasons with the temperature in winter falling to as low as –15C in the north. Economically, too, Japan may not be classified as "southern," with its GNP ranked third in the world and with its highly advanced infrastructure and relatively high living standard. If the Global South is a nomenclature used as an equivalent for another journalistic password "Third World," and if the concept deals solely with such issues as poverty and other forms of economic and political exploitation, Japan, then, may not be a suitable subject to take up.

In the following essay, however, I will redefine some of the implications of the concept of the Global South and attempt to clarify its economic as well as political ramifications. My premise is that the Global South is a symbolic as well as historical concept: it fundamentally refers to a certain stage of social development, either within individual countries or in larger geopolitical regions. In my framework of discussion, therefore, the "South" stands for the local and regional as well as the global, and it also refers to various forms of oppression, including regional inequalities and environmental injustice. Given the recent economic rise of BRIC and the rapid industrialization of Southeast Asian countries, traditional demarcations of the "South" have increasingly been rendered unaccountable, and the locus of social issues nowadays tends to become more local and regional in many countries.

Based upon this definition of the Global South, I will discuss the case of Minamata Disease, the most extensive and disastrous industrial pollution that broke out in the period of Japan's rapid postwar economic growth in the 1950s and 1960s. The text I deal with is Ishimure Michiko's *Kugai Jodo* (1969), a work of creative nonfiction which, on account of its profoundly evocative power, has exerted a tremendous influence and actually determined the course of the protest movement that still persists today after sixty intervening years.

A. MINAMATA AND ITS AFTERMATH

A younger generation of Japanese today, gravely concerned about the health risks of PM 2. 5 (particulate matter 2.5 micrometers) carried in the wind from China, would be a little surprised to learn that Japan herself was once one of the most polluted nations in the world and a major producer of pollutants in the air and the surrounding waters. During and after the period of unprecedented economic growth in the 1950s and 1960s, the country found itself suddenly confronted with the deadly outcomes of rapid industrial expansion: the sky was shadowed even in the midday by complications of industrial fumes and other emissions, with occasional emergences of photochemical smog. Along coastal areas that had a heavy concentration of factories and oil plants, several feet of pollutant-filled sludge accumulated at the bottom of the sea, thereby totally devastating the rich ecosystem of the tidelands.

In view of the rapidity with which the natural environment was mercilessly destroyed, two social and industrial factors would perhaps stand out as its major causes. One was the quick progress and expansion of the chemical industry, and the other was the total lack of regulatory measures and monitoring systems, which resulted in the unharnessed drainage of wastewater into rivers and the sea. By the time the national environmental agency and the first anti-pollution act were established, respectively in 1971 and 1972, lawsuits had already taken place over the loss of life and compensation liabilities. Among the cases sensationally publicized by the media were the four major lawsuits involving the incidents of industrial pollution: Minamata Disease, Niigata Minamata Disease, Itai-Itai Disease, and the Yokkaichi Asthma Case.

Minamata, located in the prefectural border of Kumamoto and Kagoshima on the southwestern island of Kyushu, is a quiet little town facing the bay, embosomed in the larger expanse of the Shiranui Inland Sea. Minamata Disease is a neurological disorder caused by the intake of organic mercury, namely methylmercury, bioaccumulated in fish and shellfish. Methylmercury was produced in the acetaldehyde acetic acid facility of the Shin Nihon Chisso factory (commonly referred to as Chisso) where mercury sulfate was used as a catalyst, and was directly discharged into Minamata Bay for more than thirty years, a practice that started in 1938 and continued until 1968. The factory's effluents contained more than a dozen highly toxic substances, such as lead, mercury, manganese, arsenic, selenium, thallium, copper, and potassium oxide, a fact suggesting the unpredictable depth of ecological disaster which also ironically delayed the official identification of the causal substance of the disease (Harada 34).

Among the symptoms of Minamata Disease are ataxia, disequilibrium, tremor, concentric constriction of visual field, and impairment of gait and speech. In radical and extreme cases, patients suffer total paralysis or, reversely, spasmodic movements escalating to ballism, a state of choreal delirium with occasional utterances of beast-like howling, and they usually die in a span of a few months or even in a few weeks. The congenital form of the disease was also discovered among younger children afflicted with cerebral palsy (Harada 67–78). It was found out that methylmercury in the mother's bloodstream passes through the placenta and concentrates in the foetus.

As of March 31, 2001, 2,955 patients had been officially certified as Minamata Disease patients; a total of approximately 144.1 billion yen had been paid as compensation from the responsible company.[1] On March 29, 2010, a settlement was reached with the compensation case of 2,123 as-yet uncertified victims;[2] debates and legal cases are still continuing over the relevance of the certification criteria.[3] The potential number of victims afflicted in some way or other is estimated to be up to 100,000 or even 200,000.

Ishimure Michiko,[4] poet, novelist, and a native of Minamata herself, closely followed the outbreak of the disease and the development of legal procedures. Her approach is not that of an objective reporter; not only did she empathize deeply with the experiences and feelings of the victims, but also actively participated in the legal and political movements. In fact, she is the strongest single voice pleading for the cause; her *Kugai Jodo* (translated as *Paradise in the Sea of Sorrow* by Livia Monnet, and henceforth referred to as *Paradise* in citations) has been considered the most influential historical document of the incident. Furthermore, the author created an untold tale of the patients and villagers whose experiences, beyond the sufferings of the disease and environmental destruction, truthfully mirror the cosmology of the native world and question the idea and value of modern civilization.

Paradise in the Sea of Sorrow, the first volume of the trilogy, was published in 1969; the revised version followed in 1972, and since then the book has gone through more than thirty reprints, reflecting its extensive popularity and enormous social and political impact.[5] The narrative focuses on the tales of four victims, Yamanaka Kuhei, Sakagami Yuki, Ezuno Mokutaro, and Sugihara Yuri, each modeled after a real life patient despite the pseudonym, and deals with a span of about ten years tracing the discovery of the disease in 1956 and the following social development surrounding the incident. The narrative, however, is by no means linear and straightforward: the chronological sequence is often disrupted by flashbacks of reminiscences, and the patients' tales are frequently cut and fragmented by the insertions of medical reports, clinical charts, excerpts from newspaper articles and official statements, and quotes from songs and poems. After the first reading, the reader may be left with a sense of disorientation, and yet simultaneously struck by the sharp edges of fragmented visions and profound depth of the suffering caused by the catastrophe.

The technique of fragmentation, which juxtaposes different types of dialects serves several purposes. These insertions are basically explanatory, providing scientific and historical information about the patients' clinical states in addition to the external social phenomena. They are also contrastive and sharply demarcate the break between the past and the present by highlighting the excruciating awareness of loss that accompanies this radical shift. The narrative also draws a distinct line between those within and those without: the language of the community, spoken with heavy southwestern dialects, is sharply opposed to the scientific and conceptual vocabulary of medical and journalistic writing.[6] The only unifying principle in the narrative, if there is one, is the narrator's viewpoint, the axis of consciousness and poetic sensibility through which the disintegration of the community is perceived and captured by way of jagged visions and conflicting languages.

B. ISHIMURE'S DISCOURSE OF THE SOUTH

Despite the fact that critics have so consistently failed to notice its true significance, Ishimure's picture of the village life, with patients' reminiscences and dreams, is strongly suggestive of the southern climate and atmosphere. Geographically, Minamata Bay and its larger surrounding region, the Shiranui Sea area, are situated in the southwestern part of Japan; and its natural climate borders on a moderately subtropical category much warmer than the rest of the country. From a wider and more global perspective, the area lies on the rim of the southern sea, connected and trafficked by the sea currents. In addition to the geographical and realistic elements, Ishimure's depiction of the seascape associates it with the discourse of the South. The sea consistently remains calm, fragrant, and bright with sunshine, often inducing fishermen to sleep; the sea is portrayed as feminine, thus giving impressions of its maternal abundance rather than paternal severity, with which the northern sea is more readily associated in Japan. The life on the sea in the narrative is a family business, husbands and wives working together for daily subsistence, and it is essentially an act of love.

Fundamentally based upon the geographical and climactic elements, Ishimure's discourse of the South is also symbolic and historical; her depiction centers around the fundamental dualisms characterising postwar Japanese society—rural and urban, natural and industrial, subconsciously animistic and scientifically rational. By emphasizing the natural and deeply animistic view of the world that the villagers inhabit, Ishimure succeeds in creating a narrative of the symbolic "South," a tale of life on the sea whose resonance is extended and elevated to a mythical sphere and paradoxically enhanced by the juxtapositions of modernizing forces. Ishimure's symbolic "South," on the one hand, reflects historically Japan's social structure in the postwar era, and on the other, it influences the narrator's self-identification with the patients, making her a crusader for their cause against the victimizing society.

The narrative of *Kugai Jodo* consists of a series of contrasts, both structural and stylistic, divided between scenes before and after the breakout of the disease. The juxtaposition of victims' tales and medical reports, for instance, puts into relief the discrepancy in the textures of reality constructed by the types of language people use. Ishimure, in other words, is experimentally engaged in building the completely opposed visions of reality focused on the physicality of the patients' existence. One is focused on life in sympathy with the natural climate, the other on definitions of death and the narrow limits of the academic discipline. Here the author fully exploits the capacities of Japanese writing systems: in delineating the patients' tales and reminiscences, she engages *hiragana,* a colloquial and expressive form, as opposed to *kanji* (Chinese characters) used for the conceptual and scientific vocabulary of clinical charts. More strikingly significant is the use of *katakana,* a form commonly used for foreign words, in referring to disinterested and even coldly cruel voices.[7]

These two visions of reality are respectively cast into reminiscences of the life at sea on the one hand, and into horrific scenes of the university hospital on the other. The fishermen's activities at sea are so visually and heroically represented that the descriptions seem to border on the mythopoetic; it is indeed in this mythic quality that we locate the symbolic "South" of Ishimure's narrative. Directly opposed to the somewhat idealized visions of the life at sea is the cold and dreary landscape of the university hospital where the patients felt "like animals in a cage" (*Paradise* 156). The patients themselves instinctively sense the horror of being taken to the hospital where the tissue of reality is composed of the same scientific vocabulary used in clinical charts—a system of construction as unnatural as the chemical factory. In a lonely ward at the university hospital, Sakagami Yuki feels that "the world is breaking to pieces" (*Paradise* 148).

C. DANCE ON A FISHING BOAT

Ishimure's discourse of the south will be sharply outlined through the discussion of some of the key images and motifs that surface consistently throughout the narrative. The patients' acute sense of loss, for instance, is delineated by references to fishing boats; the boats, fishing gears, and even harbors are described as psychological and symbolic extensions of their former physical existence. Fishermen of the Shiranui Sea areas, upon arriving in Minamata harbor for mass protests, witness a cluster of abandoned boats at the deteriorated wharf, a symbolic death of the village life. One of the fishermen comments, "But this ghost harbor, damn it, it's like your own grave suddenly open up at your feet" (*Paradise* 70).

A more pathetically dramatic scene linked to the image of boats is Sakagami Yuki's reminiscences of her former life. Yuki came from the Amakusa Islands on the opposite shore of Minamata Bay and remarried Mohei. Yuki's attachment to their boat and the pastoral vision of the life on the sea are radically replaced by the scene in the University Hospital where she is clinically forced to have an abortion and envisions her demented self chasing after and devouring the discarded baby. The nurses at the University Hospital take pity on her state and make her paper boats laden with candy. Again Yuki's suffering, loneliness, and despair are recapitulated by the image of a fishing boat. "I'd walk up and down the corridors at the University Hospital, dragging a small paper boat and shouting '*essin yo! essin yo!*' as we used to when we had to haul in a net bursting with fish. My soul lay in that tiny paper boat, buried under a pile of candy" (*Paradise* 174).

Dance is another frequently used motif in the narrative in relation to the villagers' physicality. Fishermen and fisherwomen are very fond of singing and dancing; their performances on festive occasions are described even heroically as manifestations of their physical strength as well as expressions of their spiritual health. Old Sensuke was a master of "Dance of the Stick," a regional martial dance; Yuki used to sing and dance, and run around laughing with children; Satsuki, elder sister of Yamanaka Kuhei, was a star dancer. Poignantly, the heroic tales of dancing are juxtaposed to the symptoms of ballism and the spasmodic movements of the patients. At the Isolation hospital, the mother watched her "toss in bed as if she were lying on burning coals. She clutched at the air with her hands and feet like a rabid dog, foaming at the mouth. Satsuki, the girl who could outdo any young man in the village" (*Paradise* 33).

D. *BON'NO* AND *TAMASHII*

Together with the images and motifs focused on the villagers' physicality, there are some spiritual elements that characterize Ishimure's discourse of the South. Ishimure reiterates a few specialized notions in the patients' tales that constitute the core of their moral and spiritual values. One is the Buddhist concept of *bon'no*. As a Buddhist term, *bon'no* is a negative concept referring to earthly desires that, in view of emancipation from the *karma* cycle, is best to be eliminated. In secular usage, however, the concept has acquired a wider connotation that vaguely implies instinctual life force, a notion similar to Freudian *eros*. Typically, the word is used in collocation by the villagers: *bon'no ga(no) waku* [feel sympathy or longing] or *bon'no ga(no) fukai* [person of deep feelings or with the capacity to show sympathy], the phrases are thus meant to underscore emotional and ecological connectedness. Yuki feels *bon'no* for fish and creatures in the sea; and she wishes to be born human again as she has deep *bon'no*.

Another item of vocabulary strongly suggestive of animism is the word *tamashii* [soul]. As with the case of *bon'no*, the word *tamashii* is used in collocation, *tamashii ga (no) fukai* [has deep soul]. Referring to Mokutaro, his grandchild with congenital Minamata disease, Old Ezuno comments, "He may be misshapen and dumb, but his soul (*tamashii*) is unfathomably deep" (*Paradise* 192), explicitly indicating the centrality of soul in their spiritual universe. The more evidently animistic usage of *tamashii* appears in the narrative of Sugihara Yuri, a young patient in a vegetative state whom the media dubs as the "milk-drinking doll." Here, too, the mother's lingering anxiety is about the state of her daughter's soul: "There is a living spirit (*tamashii*) in every tree, in every weed, in every blade of grass. Fish and earthworms, all living beings are endowed with a soul (*tamashii*) which stays behind and enters a new life when they die. How can it be that our Yuri hasn't got a soul (*tamashii*)?" (*Paradise* 260).

The villagers' view of life, with strong emphasis on either "deep soul" or "deep *bon'no*," is characteristically opposed to the values in modern industrial society centered on the notions of individualism and scientific rationalism. One is focused on human connectedness to the organic whole, thus serving to sustain and harmonize with the natural climate, while the other centers on analytical science and the process of atomistic reduction. The language of the villagers, which uses the southwestern dialect, and their views of physical reality as being deeply imbued with spiritual meaning, directly challenge the modern way of perception as well as the validity of our communication. Minamata Disease, a byproduct of chemical reactions itself, therefore, is a symbolic and ultimate consequence of modern society with its rationalistic thinking and analytical methods. "Minamata Disease," comments the narrator, "is a fundamental issue in our time, a finger pointed accusingly, as it were, at the philosophy and morals sustaining modern civilization" (*Paradise* 239).

It is indeed the case that Ishimure's discourse of the South is greatly enhanced by her visionary capacity. Her creative vision, in point of fact, endows the narrative with rich mythical implications that collectively constitute the symbolic implication of the South. Ishimure's imaginative capacity is best demonstrated in the scene of the patients' first public demonstration in Minamata in 1968. Ishimure envisions a ritual procession of rain dance in the feudal era while participating in the protest march of a group of about hundred patients and supporters.

> We speak unto you, o Dragon God, and
> Unto you
> O gods of the sea.
> Quell your anger and listen to our prayer,
> This virgin, as beautiful as a princess of old,
> Will be thrown into the Sea.
> Give us rain, give us rain!
> Without rain, trees and grasses wither
> And men die in agony.
> Take this girl, o gods, and give us rain! (*Paradise* 335)

Here the author is not simply reiterating the regional history or native cosmology. The reference to the Dragon God, a guardian god of water, rivers, and the sea, is ironically significant in that the "anger" of the nature god is undoubtedly associated with the outbreak of Minamata Disease—a direct consequence of the contaminated seawater. It is precisely for this purpose that the author, in her imagination, revives Old Sensuke and allows him to demonstrate his masterful "Dance of the Stick" in the procession.

Another important motif here is that of human sacrifice. It is evident that a young village girl to be sacrificed in the ritual prayer ("This virgin, as beautiful as a princess of old") is taken as a prototype of Minamata patients, such as Sugihara Yuri, a "milk-drinking doll."[8] Ishimure here suggests that the Minamata patients are not simply victimized by the exploitative operations of the Chisso corporation (Chisso continued the criminal practice of discharging the polluted effluent into the sea until 1968, more than ten years after the first patient was officially recognized), but they are literally sacrificed by the rest of community which sternly ostracized them, suspecting that the disease was contagious. In a larger sense, this holds true for the nation as a whole, since it embraced the rapid economic growth and industrial progress in the postwar era. Through the viewpoint of a member of the lowest social class, and especially through the eye of the weakest victims, Ishimure, as a visionary "sorceress," reimagines the shape of the country and directly questions the nature and character of its modernization process.

Minamata has been called, rather ironically, the "castletown" of Chisso, suggesting that the company lords over the township. Economically, the revenue of the city has been dominated by taxes from the corporation as well as from its thousands of employees. Politically, too, the city government has been closely entwined with the factory officials: Hashimoto Hikoshichi, the twice-elected mayor, was the leading engineer of Chisso and the factory manager at the end of the Pacific War. Viewed historically, village leaders of Minamata, after the collapse of the local salt industry in the 1930s, enthusiastically invited Chisso to set up its plant in their village; they saw in Noguchi Jun, the president of Chisso "the very personification of Progress" (*Paradise* 109).

In other words, Ishimure's symbolic "South" is not simply opposed to the modernity of the "North," but enjoys an ambiguous relationship with the other wherein the "South" embraces the "North" within its bosom. To demonstrate this consciousness, Ishimure juxtaposes two songs sung by children when she was a schoolgirl: one is the school song of the First Minamata Elementary School, and the other the song of the Chisso's Minamata factory. Ishimure tries to show how in the children's minds the natural climate celebrated in the former subconsciously merges into the modern principle of progress in the latter. The people of Minamata, who learned these two songs by heart, would recall that the lines "Curly smoke mingling with white clouds / Lingering above the town," in particular, did arouse a fresh and healthy sense of pride in the rising new town. Just as organic mercury surreptitiously enters and settles in the human body, so too the chemical plant of Chisso has been assimilated in the rich southwestern climate of Minamata and embedded in the intimate relationship.

In the section titled "My Homeland and the History of the Factory," the author delves into the history of the Chisso Corporation and tries to illustrate some characteristic features of modernist ideology. Significantly, she draws on her childhood experiences relating to her demented grandmother who was ordered to be isolated on an island due to the Emperor's visit to the Chisso factory—the chemical industry star. The picture of the state power closely allied to industrial capitalism, and the following logic of exclusion and isolation, basically determines the framework of perspective through which Ishimure examines the modernization of Japanese society, with the Minamata Disease issue as its microcosm.

The rapid growth and expansion of Chisso is not solely dependent on technological progress but closely allied to military aggression and the history of colonization. While the Korean peninsula was under Japanese colonial rule, Chisso established a branch factory in the coastal village of Honam in Hamgyong Nambo Province in 1927; Noguchi, the president, set out to purchase the exclusive rights over Changjin River in 1935, a policy that was strongly supported by Ugaki Kazushige, the governor-general of Korea. Thus, Chisso's great strides in the chemical industry were largely a result of "the 'glorious march forward' of Japanese militarism and the 'extraordinary development' of Japan's colossal, monolithic business combines" (*Paradise* 290). Inevitably, Ishimure's attention is directed toward the fate of the ostracized Korean villagers and the manner in which their native land was confiscated in the presence of the Japanese police force.

Here Ishimure clearly detects the same ideological trait or pattern of thought in Chisso's dealings with the Minamata patients and the Korean villagers. On a deeper level, the capitalist logic of progress and expansion is sustained by exploiting the lower strata of people, and it is constantly and explicitly reinforced by "the government's more or less open policy of slow annihilation of the lower class" (*Paradise* 329). This capitalist sensibility penetrates into the core of modern consciousness and permeates into the tissue of Japan's reality: "Today it reverberates in the sick condition of our country, in the corruption of morals and devaluation of human feelings which make up the dominant characteristic of our time and which are also my hateful inheritance" (*Paradise* 293).

In the "Author's Afterword," Ishimure states her creative intention by reiterating "the contrast between the beautiful character of the Minamata Disease patients and the murderous logic of industrial corporation" (*Paradise* 375). Against the deepening shadows of the capitalist logic and those faceless multitudes subserviently dependent upon it, the Minamata patients' spirit of forbearance and independence increases in radiance. Ishimure's discourse of the "South" derives essentially from the natural climate of southwestern Japan with its abundance of sunlight, and yet symbolically focuses on the beauty and radiance of those who in spite of having suffered beyond the limits of human endurance are still capable of a greater sense of identification with the world around them.

WORKS CITED

"Agreement reached to settle Minamata suit," *Asahi Shimbun* news, 31 March 2010, Web. 5 August 2013.

Harda, Masazumi. *Minamata Disease*. Trans. Tsuhima Sachie and Timothy George. Kumamoto: Kumamoto Nichinichi Shinbun, 2004. Print.

Ishimure, Michiko. *Kugai Jodo: Waga Minamatabyo*. Vol. 2 of *Ishimure Michiko Zenshu* [Ishimure Michiko: Collected Works]. Tokyo: Fujiwara Shoten, 2004. Print.

———. *Hi no Kanashimi* [Sadness of the Sun]. Tokyo: Asahi Shimbun, 1986. Print.

Kawamura, Minato. "Kaze o Yomu, Mizu ni Kaku" [Read the Wind and Write on Water]. *Gunzo*. June, 1997: 384–404. Print.

"Minamata Disease: The History and Measures," The Ministry of Environment (2002), Web. 5 August 2013.

Monnet, Livia. *Paradise in the Sea of Sorrow: Our Minamata Disease*. Kyoto: Yamaguchi Shoten, 1990. Print.

Takahashi, Tsutomu. "Kotoba no Kindai: Ishimure Michiko niokeru Bungaku to Fudo" [Language and Soul in the Works of Michiko Ishimure]. *Literature and Environment: The Journal of the Association for the Study of Literature and Environment in Japan* 6 (2003): 30–37. Print.

———, Sadamichi Kato, and Reiko Akamine. "The Conservation Movement and Its Literature in Japan." *Literature of Nature: An International Sourcebook*. Ed. Patrick D. Murphy. Chicago: Fitzroy Dearborn, 1998. 290–93. Print.

———, and Noda Ken'ichi. "Interview with Ishimure Michiko." Vol. 16 of *Ishimure Michiko Zenshu*. Tokyo: Fujiwara Shoten, 2013. 617–33. Print.

Watanabe, Kyoji. "Ishimure Michiko no Sekai" [The World of Ishimure Michiko]. *Kugai Jodo: Waga Minamata Byo*. Tokyo: Shinchosha, 1972. 305–25. Print.

NOTES

1. "Minamata Disease: The History and Measures" (http://www.env.go.jp/en/chemi/hs/minamata2002), The Ministry of Environment (2002), retrieved 5 August 2013.

2. "Agreement reached to settle Minamata suit" (http://www.asahi.com/english/TKY201003300438.html), *Asahi Shimbun* news, 31 March 2010, retrieved 5 August 2013.

3. Most recently, the verdict of a legal case involving the certification criteria came out in May 2013.

4. Japanese names in this essay are spelled with surnames first as is the customary usage in Japanese writings. In this case, Ishimure is the surname; Michiko is her first name.

5. Livia Monnet, in "Translator's Introduction" to her *Paradise in the Sea of Sorrow*, comments on the book's enormous social impact: not only did it make way into the part of the literature curriculum in high schools, but it was also adapted for TV, made into dramas, and used as lyrics for blues ballads. Livia Monnet, trans. *Paradise in the Sea of Sorrow* (Kyoto: Yamaguchi Shoten, 1990) iv.

6. Ishimure's stylistic experiment has been commented on and studied by a number of writers and scholars. See Kawamura Minato, "Kaze o Yomu, Mizu ni Kaku," *Gunzo* (June, 1997): 384–404; Watanabe Kyoji, "Ishimure Michiko no Sekai," in *Kugai Jodo: Waga Minamata Byo* (Tokyo: Shinchosha, 1972) 305–25; Takahashi Tsutomu, "Kotoba no Kindai" [Modernity of Words], *Literature and Environment: The Journal of the Association for the Study of Literature and Environment in Japan*, 6 (2003): 30–37.

7. Takahashi points out the importance of *katakana* in *Kugai Jodo*, in comparison with one of Miyazawa Kenji's short stories, "Kun Nezumi" [Kun Mouse]. See Takahashi, "Kotoba no Kindai."

8. The motif of young girls victimized is further associated in the narrative with the instance of sacrifice historically determined by the socio-economic situations of the region: the practice of human trafficking. Young daughters of impoverished peasant families were sold to brothels either within or outside the country. The practice had been conducted not only in cruder societies in the feudal era, but it also persisted even in the modern times. Some of those young daughters were sent overseas to China, the Philippines, Malaysia, and Indonesia, thus connecting the issues of poverty and oppression with the case of southern diaspora.

FIVE

Indian Environmentalism and its Fragments[1]

Jyotirmaya Tripathy

A. INTRODUCTION

Ever since Gadgil and Guha's intervention in India's environmental history and their subsequent division of people into three categories—namely the omnivores, ecosystem people, and ecological refugees—it has become rather customary to see Indian environmental movements as "environmentalism of the poor" which is built "on the question of equity" (98). The implication is that environmentalism in the industrialized North is predominantly elitist and thus is a rational response to a global crisis. Third World or Southern environmentalism, in contrast, is concerned with questions of equity and distributive justice, where environmental awareness is only incidental and secondary to the imperatives of life and livelihood. This model leads to an interesting form of solidarity that connects the poor and underdeveloped of the globe struggling for survival and pits them against the rapacious nature of the capitalist system. Third World environmentalism thus grows not just as a counter narrative to global capitalism and its negative externalities vis-à-vis environment and the poor, but also offers a critique of the perceived apolitical and disengaged nature of First World environmentalism.

When Third World environmental difference vis-à-vis the First World is a taken-for-granted reality, it cannot be isolated from the idea of the modern postcolonial state which so far has been "catching up" with the West. This quest often comes in conflict with the First World environmental consciousness which seeks to rescue Third World ecology from degradation as the postcolonial state sanctions the plunder of its hills, rivers, and forests to satisfy its increasing resource needs. The postcolonial state legitimizes its resolve to over-exploit the environment through the narrative of development which gives it immunity not just from the international community, but also from its own people. Its disregard for the environment comes from the West's duplicity in terms of its past plunder of resources, and also from the aspiration of Third World countries who want to approximate the West regardless of its effect on the environment.

While it is true that Third World environmental activism is a response to developmental challenges posed by the state and the penetration of global capital into every nook and corner of the world, there is an uncritical and unproblematic theorization of such activism that often reduces environmentalism to questions of life and livelihood. When the question of equity is foregrounded and projected as self-evident truth, when the political origin of such mobilization is traced to livelihood, and when the articulation of this resistance is seen as uniform, what is ignored in the debate is the contested and multidimensional nature of such activism, its mediation through other categories like caste and gender, its multiple and often contradictory expressions, and the unevenness in the delivery of such activism. It will be more productive to ask questions such as: How do competing and conflicting conceptions of environment and development between First World and Third World (and also within Third World) bring forth new dimensions to human-environment relationship? Which facets are foregrounded when we produce particular environmental narratives and what do they elide? How do the political expressions of these movements repress and produce resistant narratives? How does the articulation of these movements produce slippages? How do activists take interest in tribal/rural landscape as an experiment with otherness and how does their social construction of geography come into conflict with affected people? What counts as environment and environmental problems?

The present paper is intended to challenge conventional assumptions of Third World environmental activism. It highlights the fractures within the practice of such activism which has implications for complicating issues like people, development, resistance, etc., and also has ramifications for questions like state and its legitimacy. The first part of the paper exposes the politics of confining Third World environmentalism to the question of livelihood and the subsequent exclusion of dissenting voices. The second part problematizes the notion of "people," leading to a more nuanced engagement; and the third part shows the production of fractures within articulations of resistance by environmental movements. The conclusion sums up the discussion, while proposing the need for addressing environmental concerns alongside other social categories of analysis.

B. THE ENVIRONMENTAL DISCOURSE

Livelihood as raison d'être of Third World environmentalism is a trope in environmental discourses as critic after critic highlights how such actions "stem primarily from a concern with livelihood" and are concerned with "productive use of natural resources" (Ghai and Vivian 1). This Third World difference has some kind of ontological solidity that forms part of any environmental scholarship passing off as an article of faith. However, such theorizing creates a polar difference between the idyllic people of subsistence and sustainability and the agents of the plundering market. If the First World dismissal of community management in Hardin's theory of Tragedy of Commons (1968) is simplistic, equally simplistic is the reverse argument that all community managed/owned resources are sustainable, which scholars like Gomes seem to believe: "Indigenous peoples are generally noted for their traditional ecological knowledges and practices that enable them to live sustainably with the natural environment" (Gomes 1065).

The centrality of livelihood in representations of the Third World projects ecological Marxism as the most important framework for responding to environmental problems. Thus political ecology research in the Third World is a response to the perceived apolitical nature of mainstream environmental literature which leads Raymond Bryant to claim that the "Third World's environmental problems are not simply a reflection of policy failure . . . but rather are manifestation of broader political and economic forces associated notably with the global spread of capitalism" (8). Though this engagement may apparently look political, its obsession with livelihood issues confines the Third World poor to a pre-political condition. Subordination of environmental awareness to the question of distribution also confines Third World activism to the pre-history of environmental consciousness. It is interesting to note that in environmental discourses, First World remains the subject of environmental history where the exploitation of environment is no longer seen as an imperative for growth given that Western countries are moving away from industrialization to service sector. The South in turn remains stuck in either pre-modern subsistence condition of the people or the mass modernization and industrialization drive by the state, but lives in the infancy of economic development.

If First World environmental discourses relegate the Third World to the pre-history of environmental consciousness and activism, Third World environmental scholars internalize such a location by placing Third World activism in the question of survival. The latter not only reproduces First World narratives of Third World hunger and poverty, but also sustains those conditions by continuously producing these narratives. When the postcolonial developmental state over-believes in the universality of Western development model and tries to play the catching-up game, Third World scholars and activists, like their counterparts in the West, see the site of environmental activism as a problem of livelihood, thus denying agency to Third World people. In statist discourses these people appear as obstacles in the path of development and so become complicit in their own underdevelopment and economic marginalization. In environmental discourses they appear as the last men standing between the environment and its destruction and are represented as ecosystem people living outside the exchange economy and content in their world of environmental plentitude until the state decides to "develop" them or acts as the agent of foreign capital. Their life, seen as uncorrupted by materialist and corroding influence of capital, is often cast as an antidote to growing environmental degradation.

Both the state and environmental discourses rob Third World people of their agency to write their own histories. The only agency we see in people is their ability to resist the postcolonial state or corporate bodies who plunder natural resources or displace them from their ancestral land. This is almost a precondition for environmental consciousness to emerge, where the authentic people are the ones betrayed by their state and who in turn should mobilize resistance with the help of national and international environmental organizations. It is these organizations and their activists who find a voice for these affected people and insert them into the web of environmental narrative, thereby rendering them subjects. Subjecthood thus does not come with the free will of the affected people, but with their subjection to a particular type of resistant discourse.

This also fits well into the over-all narrative that people are fighting for survival. Who can be against a movement which guarantees or at least fights for the survival of its members? Metaphors of livelihood and survival, unlike other categories such as race or class, are not questions of ideologies but something very basic and foundational—the question of life. It is this obsession with life which reduces Third World environmentalism to a mere biological condition and Third World people to the status of zoe.[2] The conflict between state and affected people then is represented as a conflict between two modes of economic principles, one of mere survival, subsistence, and harmony with nature, and the other guided by market capitalism, profit, greed, and exploitative relations with nature. The difference is highlighted to such an extant that these movements are seen as a fight between good and evil.

Mention may be made here of an environmental/livelihood movement in the eastern Indian state of Odisha The South Korean steel major POSCO entered into an agreement with the state of Odisha in 2005 to build a steel plant over 4000 acres of land in Jagatsinghpur district. Since then the affected people have been fighting against the land acquisition by the state administration. We have POSCO Pratirodh Sangram Samiti (PPSS), roughly translated as Association against POSCO, which has been at the forefront of the movement demanding the withdrawal of the Korean company. There are also fifty-two families of POSCO sympathizers who were believed to have betrayed the cause of the movement, the reason they were forced out of their village by PPSS. But in my interaction with some of these displaced people it emerged that these fifty-two families were actually against POSCO in the beginning. However, on account of being Congress party workers, they were often outnumbered and challenged by the CPI (The Communist Party of India) cadre in their village.

It is the CPI cadre who will later form the PPSS and drive away these Congress workers from the village after some petty problems over panchayat elections flared up. Much before the anti-POSCO agitation began, these families had already been displaced by CPI cadre. Ironically these families were branded pro-POSCO because they depended on the district administration for their safety on a regular basis which had nothing to do with the fact that administration was acquiring land. Such is the power of environmental discourse and such is its authority to confer acceptability that anybody outside its narrative is represented as an apostate, which leads to the loss of his land and home. This PPSS type environmental activism thus produces a new kind of citizenship which is oppositional to the state and which derives its legitimacy from that resistant subject position, often decided by NGOs, Communist parties, Third World intellectuals, or its own leaders. Though subjection to the state is seen as loss of agency in this narrative, subjection to these resistant environmental narratives interpellates development affected people and environmental activists as Subjects.

There is complete indifference in various environmental media about these fifty-two families who were forced to leave their village in 2007 and later agreed to the terms and conditions of land acquisition by the state. This throws up an interesting facet of environmental politics which usually questions foreign-invested and state-facilitated development projects which people think will impoverish them, displace them, and in the process destroy their livelihood, epistemologies, land ethics, and meaning in their lives. What is ignored is the emergence of another repressive narrative which cannot tolerate any dissent. Scholars of the environmental movement often ignore the plight of these people as their number is insignificant, and also because their narrative does not fit into Third World difference and their so-called alternative paradigms of development. By their act of support to "development" projects, they destabilize the conventional notions of subsistence farmers and environmental subjectivity.

C. ENVIRONMENTAL SUBJECTIVITY

The most interesting term used in these environmental discourses is "people" who form the foundation of these discourses, give them their legitimacy, and produce environmental subjectivity. These people of environment are seen as an unchanging mass outside exchange economies, closer to nature or living a life outside the predatorial nature of state and global capital. They are also seen as uniform, sharing precisely the same interest in the environment. If there are any conflicts relating to class, caste, etc., people are believed to transcend these irritants for the greater cause of environment. As Amita Baviskar says, for the movement to be successful, it "must convert fluid identities into primordial and unitary categories: tribes, women, or peasants; all must appear to be internally undifferentiated and uniformly opposed to the state" (*India's Environmental History* 520).

These people are also represented as living like one family until some Foreign Direct Investment (FDI) or development plan threatens their survival. All along they are territorialized and seen as belonging to the land; the only identity they have is their concern for the environment. Either they are infantalized in terms of their concentration on livelihood, physical well-being, and survival, or else they are romanticized as people of nature. But the environmental subject here is like the child living in this world of innocence, uncorrupted by capitalist thinking that exploits the environment. Since the child cannot sever ties with the environment-mother, the castration threat from the state or global capital is a traumatic experience for these people. The native here is the permanent child of the Oedipus complex or the semiotic order. This world of capital is the world of symbolic/patriarchal order where the child cannot speak, or at least cannot speak in a language that the state can understand.

Often the literature on these environmental movements is celebratory in nature offering uncritical views of people and their attitude towards nature. Dunu Roy exposes the uncritical references, either in environmental or political discourses, to "people" as a mass waiting to be helped and guided: "The notion of 'people'—so attractive to environmental discourse—lay at the core of these political statements" (526). Roy questions this simplistic notion of people, as there is no consensus as to what it meant other than some vague notion of dalits, women, tribals, and backwards. It comes as no surprise that many environmental action groups valorise the idea of "people" and highlight their centrality in their names. Often they are called People's Action Group, People for Land Rights, People's Union for Human Rights, etc., which gives some kind of representative character to their actions.

What is conveniently ignored is that the term "people" is slippery to say the least and often carries more contradictions within it than solidarity. Though the term evokes automatic legitimacy in a democratic state like India, it will be a little radical to say that the term's legitimacy comes from a framework which is statist in nature and against which most of the environmental activism is directed. Thus it is no surprise that the Indian state often uses the same vocabulary (like "inclusive growth" or "people-centric approaches") as environmental action groups because of its representative character. Both the state and environmental action groups draw legitimacy from the public; it is just that their conceptions of "people" are different. For the state, "people" may refer to any group, real or imaginary, or even the state which will get benefit from a project. In contrast, for environmental groups, the term may refer to inhabitants of a particular area who may be negatively affected. It is this competing notion of "people" which remains at the heart of conflict between state and environment groups.

Homogenization of people is driven by distance; it is the latter which sees unity within fragments and uniformity within conflict. Like all mobilizations, environmental action is not immune to various sub-interest groups working within it and jostling for voice and representation. But environmental discourses, whether local or global, usually minimize differences within the group and maximize cohesion. Here Seyla Benhabib's (2002) intervention may be rewarding when she distinguishes between social observer and the social agent. The researcher or the external activist usually sees the environmental movement as a complete formation where there is no fracture and no contradiction; everybody thinks and speaks in the same voice. This observer always sees unity and homogeneity which is an attempt to discipline the movement so that it can be a classic case of Third World movement. As Benhabib says, "from within, a culture need not appear as a whole; rather, it forms a horizon that recedes each time one approaches it" (5). It is not just academic discourse, but also environmental leadership, that does not recognize these contradictions.

This notion of a uniform people obfuscates our understanding of social life and their production of environmental knowledge. Environmental movements are often represented as a container of a group's politics where everybody responds to development policies the same way. Just how flimsy this question of community is will be explained through another environmental movement in Odisha. The Dongaria Kandha tribes of Odisha are up against Vedanta Aluminium Limited (VAL) which has the right to mine bauxite from the Niyamagiri Hills, as they are opposed to the construction of a refinery in Lanjigarh. It may be noted here that the Niyamagiri Hill is worshipped as God and is named Niyamaraja or Budharaja by the Dongria Kandhas.[3] This perfectly suits the global environmental discourses of tribals living in the lap of nature with a premodern and pre-capitalist epistemology. The organization at the forefront of movement against Vedanta Resources is Niyamagiri Suraksha Samiti (NSS) or Association for the Protection of Niyamagiri. The movement reaches the corridors of power in Delhi through lobbying and also in London where activists sensitize the aggrandizing mentality of the company and its mission of uprooting tribals from their place.

What is overlooked is the existence of another group called Lanjigarh Anchalika Vikas Parishad (Association for the Regional Development of Lanjigarh), a pro-"development" organization which accuses NSS of stalling the progress of the region. One of the leaders of this group is Jitu Jakesika, who was once actively associated with the NSS, but now claims that he was misled by NGO activists who did not want his area to develop. In the mainstream discourses, though, people like Jitu are often seen as traitors or as company agents who have sold themselves for money or alcohol. While it is easy to question state's idea of progress or development, what is elided is the question of authenticity and representation. It is more or less taken for granted that it is the anti-industry group which is the authentic representative of Dongaria Kandhas because they conform to our preconceived idea of a tribal. When the state or company represent pro-project activists as authentic tribals, NGOs and other environmental groups dismiss them as a small faction. When the first group is represented as the good guys with a belief system of their own and as interested in survival, the latter group is often represented as morally and materially corrupt who have betrayed the cause and the very core of tribalism.

So far as POSCO project is concerned, it originally required 4000 acres (later reduced to 2700 acres) of land in grampanchayats like Dhinkia, Nuagaon, and Gadakujanga. Dhinkia grampanchayat, consisting of four villages like Dhinkia, Gobindpur, Trilochanpur, and Patana, has been the ground zero of this movement as the PPSS leader, Mr. Abhay Sahoo, hails from this area, though Nuagaon and Gadakujanga grampanchayats where land acquisition has been more or less completed have by now almost succumbed to the company and administration. Every now and then the movement gets support from noted human rights groups and environmentalists. Medha Patekar and Swami Agnivesh too have lent their support to the movement. Similarly, People's Union of Civil Liberties (PUCL) members visited the site and saw large scale rights violation. Various national level organizations like PUCL and Environment Support Group (ESG) set up fact-finding teams and conducted seminars on state repression in Dhinkia. These events are exhibited in international fora like the Portugal-based *Alice* or ethicsinaction.asia. Indian forums like countercurrents.org, citizen-news.org, socialism.in also highlight state repression and the people's resolve to resist.

USA-based National Public Radio sees the anti-POSCO struggle as a struggle between tradition and modernity. Corey Flintoff, writing for NPR argues that "plans for major development projects come into conflict with traditional ways of life centred around farming" (Flintoff). Such mentality springs from the conviction that agriculture in the Third World always promotes diversity. Vandana Shiva, the famous anti-globalization activist, visited Gobindpur village and offered support to the agitation and reportedly said that "There is no justification for setting up the steel plant in the fertile land in the area where rural economy has prospered with multi-crop farming and other economic activities like fisheries and plantation" ("Vandana Shiva Joins Dharna against POSCO Project"). This goes well with her advocacy of multi-crop farming through indigenous seeds and techniques. But what is interesting is that in these areas farmers are predominantly into cash-crop farming like betel vines which are grown on government or forest land. Shiva also justified the use of children in the movement (after being moved out of their schools to participate in the movement) and argued that it involves their existence and future. These voices often see development as top-driven projects and fail to see conflicting notions of development among the people affected, imagining that state-driven development is always anti-people.

D. ENVIRONMENTAL SLIPPAGES

Environmental groups in the Third World are apparently interested in the question of survival and hence are interest groups. This may be unlike environmental activism in the West where people look at environmental protection almost as a matter of faith, and so may be called identity groups. Amy Gutmann argues that "identity group politics is bound up with a sense of who people are, while interest groups politics is bound up with a sense of what people want" (15). I would like to argue that every environmental group may have some sense of collective interest in the beginning of its mobilization, but as the movement progresses, as the negotiation starts, as the question of leadership comes, these interest groups throw open social fractures which were invisible in the beginning. Ignoring these conflicting versions of development and environment and other social categories like class, caste, and gender is a kind of epistemic violence to discipline the movement in a particular way. The moment the movement appears to crystallize around some key concerns, other concerns emerge at almost the same point of time which have the potential to render the movement as not-so-inclusive. It may also be added here that some environmental action groups are simultaneously identity and interest groups when they concern both questions of livelihood and cultural difference. When a hill or river can affect a tribe's physical and economic well-being, it can also compromise their sense of culture and ethnicity of which a river or a hill is an integral part (as in anti-Vedanta protest) and displacement will lead to loss of identity.

It is vis-à-vis the mighty state, which often transgresses its limits of control, that resistance to development re-presents itself as environmental movements. This occurs within a template of a global discourse that produces an alternative legitimacy on the basis of its mainstreaming people as the loci of development. Amita Baviskar exposes the interpellation of such movements into global discourse when she says that "claims to environmentalism are more likely to be accepted if they fit into a pre-existing template of 'green politics' as developed in the North" (*India's Environmental History* 519). Thus environmental citizenship comes with the subjection of a version of livelihood question to global vocabulary which not only co-opts this Third World problem, but also produces this as an environmental problem. This complicates the very idea of the spontaneity of Third World environmental movements, and also the very existence of such movements outside its appropriation by the West. If these movements are to be known as environmental movements, it will be made possible only through production of narratives that speak of ancestry, continuity, primordialism, and spiritual kinship with environment. It is with this sustained production of tribes or peasants that Third World environmentalism will gather its strength and will be part of the global discourse.

In reality this is more complex than what mainstream environmental narratives make us believe. During my visit to the POSCO project area, I discovered to my discomfort that people of Dhinkia grampanchayat are now slowly reconciling to the reality of POSCO and now are interested in getting maximum compensation from the administration. It may be noted here that the earlier compensation package of Rs. 4500 per square foot of betel vines has been revised to more than Rs. 10000. It also came as a surprise to me that in the hope for compensation, some smart villagers are putting together makeshift betel vines overnight so that they can be razed and compensated by the administration. Here is something that complicates the idea of ecosystem people in a sustainable relationship with their environment. The administration also pays for each tree that is cut, in spite of the fact that these are grown on government land. In the absence of clear titles over land, trees are often claimed by more than one owner to get compensation of Rs. 1500 for each tree. Similarly when the company announced one job per family, many families divided themselves by obtaining separate ration cards for each son or brother so that they can pocket as many jobs possible.

It may be noted here that both the protesters and supporters of POSCO believed in the development of their region; it is just that their notions of development are conflictual. The former President of Joint Action Committee (formed by the fifty-two displaced families) Mr. Anadi Charan Rout argues that POSCO is good for the region as it will create jobs and provide economic stability. Here he may appear to be speaking the language of the state and statist notion of development. But it will be simplistic to confine him to statist framework as he also speaks the language of life and livelihood like anti-POSCO activists. Rout reasoned that the water table of the region is decreasing and it is becoming increasingly difficult to earn a substantial living from agriculture. Even farmers with betel vines are affected because of changing soil conditions. For him long-term solutions are required, and that means industrialization. But the language of development these JAC people speak is not something that contributes to Third World difference; it also destabilizes the secure notion of the epistemological, cultural, and material difference of the Third World. Environmental activists would much rather see these people as part of the landscape, as people in their places, the defenders of pre-modern subsistence farming.

Amita Baviskar has done interesting work in this regard vis-à-vis Indian environmental movements and proposes that "the dynamics of tribal politics often strain the discourse of environmentalism and have an ambivalent attitude toward development" ("Tribal Politics" 289). Far from being a spontaneous outburst against state/corporate-led development projects, there are conflicts between activists (many of whom come from urban areas with good education and dream of being agents of change) and tribal leaders within the movement. While the former are fired by environmental consciousness and sustainability, which are often taught in their university departments, and driven by a desire to protest state atrocities and rescue the tribals from the state, the tribal leaders are interested not just in the question of environment, but also other issues like distribution, representation, and recognition. Many times they respond to these problems alongside environment, thus making the latter a multi-faceted and multi-dimensional movement, unlike external activists for whom the environmental question is of paramount importance. This tangles the unitary and one-dimensional versions of environmental movement. This also minimizes fractures or other categories like identity and ethnicity and maximizes shared norms and tries to make the movement appear uniform. Recognizing these conflicting layers of affiliation and belonging, Baviskar argues that "the discourse of environmentalism is not consistent or finished but consists of the interweaving of often contradictory political thought and action" ("Tribal Politics" 289).

This tension between environmental activists and dalits is highlighted by Gail Omvedt. She sees dissonance between these two social movements and narrates how she came to see this first hand in an NGO conference when one dalit activist highlighted the hypocrisy of environmental activists who live in cities and come periodically to lead environmental movements. For activists participating in environmental movements is like an internship where they can experiment with their leadership quality, get exposed to the taste of otherness, and show solidarity with class/caste brothers facing eviction. What distinguishes these people from tribals/dalits is a vast difference between material and epistemological realities that shapes the way people live their lives. Omvedt quotes an adivasi saying: "If you like the forests so much, why don't you come and live here and give us your city flats?" (12). Though Omvedt is highlighting an important part of the insider/outsider debate and the question of authenticity, her simplistic view cannot answer as to why adivasis and dalits are fighting for their land and forest rights. But it does raise valid questions about environmentalism as an ideology which is often transplanted and does not always come from within.

Baviskar also brings into light the ambivalence of tribal leaders toward tribal-nature relationship which activists believe is the bedrock of tribalism. However, a tribal leader Amarsinh Chaudhari once said that though tribals have a relationship with forests and land, it is because they have no other choice and that if they could, they would come out of the forest and benefit more (Baviskar, "Tribal Politics" 311). We may say that the tribal-nature connection is a First World (or Third World academic) invention sustained far too long to produce knowledge. In such a scenario, there is a subtle attempt to produce tribals in environmental discourse that conforms to global ideas of Third World activism, though it elides the complex relations tribals actually have with the land and also the larger question of development. Though tribals are silenced by academics/activists and have become objects of knowledge, they interestingly have the privilege of being the topics of environmental debate. But while enjoying the epistemic privilege of the oppressed, they simultaneously are dispossessed from authoring their own stories.

E. CONCLUSION

Every social movement, or any social formation for that matter, is not a seamless history of belonging, but is a site of contradictions, multiple positions, and factionalism. It will be academically profitable to engage with not only the official articulations of a movement, but also the voices that undo, reorient, or dilute the official versions. These hidden transcripts will contest and disturb not just the state version of development and progress, but also the counter-narratives to state development finding expression in environmental movements. These declarations may not corroborate statist developmental reason nor can they dislodge environmental movements as counter narratives, but they problematize the simplistic nature of Third World environmental discourse and its political manifestations. These transcripts are often slippery as they escape systematic mobilization, fail to codify their political positions, and establish knowledge of their own, but they betray the slippages of environmental movements. These articulations also are often without known authorship, and even if they do not disturb environmental movement itself, they highlight its inner contradictions.

Of late, Third World environmentalism has received attention from academia, mostly from scholars associated with postcolonial studies, though the field of postcolonial ecocriticism has repeated the conventional wisdom that environmentalism in the South is a "critique of current globalizing practices that use capitalist ideologies of development to justify corporate expansionism" (Huggan 6). Though the postcolonial project is laudable for liberating environment from the epistemological violence of First World environmental discourse and its tutelage, it often foregrounds the equity question, thereby exaggerating Third World difference. Though it is true that "our perception of nature is mediated by race, class, gender, and geography" (Feldman and Hsu 200), it is equally true that as categories of analysis, none of these terms is always capable of carrying the politics of the movement, as every now and then other categories resist one another while denying each category the power to contain the movement. Even when academic literature has been slowly rising to see these emerging realities, environmental activism is struggling to cope with alternatives challenging its unified narrative. Once these fragments are recognized, it will make environmental movements more inclusive and democratic. This will lead to proper politicization of environmental discourse that acknowledges multiple understandings of development and livelihood within the broader framework of the environmental movement, and not simply in the latter's resistance to state and capital.

ACKNOWLEDGMENT

The paper has benefitted immensely from my interaction with Dr. Dharmabrata Mohapatra of Ravenshaw University, Odisha, who helped me in arranging trips to POSCO project area and in decoding what various people of that area said, meant, or implied. I also acknowledge the generosity of Dr. Swarnalatha for her supply of reading materials and flexible time line to complete the paper.

WORKS CITED

Agamben, Giorgio. *Homo Sacer: Sovereign Power and Bare Life*. Stanford: Stanford UP, 1998. Print.

Baviskar, Amita. "Tribal Politics and Discourses of Indian Environmentalism." *Nature in the Global South: Environmental Projects in South and South-East Asia*. Ed. Paul Greenough and Anna Lowenhaupt Tsing. New Delhi: Orient Longman, 2004. 289–318. Print.

———. "Written on the Body, Written on the Land: Violence and Environmental Struggles in Central India." *India's Environmental History: Colonialism, Modernity and the Nation*. Ed. Mahesh Rangarajan and K. Sivaramakrishnan. New Delhi: Permanent Black, 2012. 517–49. Print.

Benhabib, Seyla. *The Claims of Culture: Equality and Diversity in the Global Era*. Princeton: Princeton UP, 2002. Print.

Bryant, Raymond L. "Beyond the Impasse: The Power of Political Ecology in Third World Environmental Research." *Area* 29.1 (1997): 5–19. Print.

Feldman, Mark B., and Hsuan L. Hsu, "Introduction: Race, Environment, and Representation." *Discourse* 29. 2/3 (2007): 199–214. Print.

Flintoff, Corey. "India's Economic Battle: Development Vs. Tradition." *National Public Radio*. Natl. Public Radio, 3 January 2012. Web. 19 June 2013.
Gadgil, Madhav, and Ramachandra Guha. "Ecology and Equity." *The Use and Abuse of Nature*. New Delhi: Oxford UP, 2004. 1–213. Print.
Ghai, Dharam, and Jessica M. Vivian. "Introduction." *Grassroots Environmental Action: People's Participation in Sustainable Development*. Ed. Dharam Ghai and Jessica M. Vivian. London: Routledge, 1992. 1–19. Print.
Gomes, Alberto. "Alter-Native 'Development': Indigenous Forms of Social Ecology." *Third World Quarterly* 33. 6 (2012): 1059–73. Print.
Gutmann, Amy. *Identity in Democracy*. Princeton: Princeton UP, 2003. Print.
Hardin, Garrett. "The Tragedy of the Commons." *Science* 162 (1968): 1243–48. Print.
Huggan, Graham. "Postcolonial Ecocriticism and the Limits of Green Romanticism." *Journal of Postcolonial Writing* 45.1(2009): 3–14. Print.
Omvedt, Gail. "Why Dalits Dislike Environmentalists?" *The Hindu* 24 June 1997: 12. Print.
Roy, Dunu. "Environmentalism and Political Economy." *Environmental Issues in India: A Reader*. Ed. Mahesh Rangarajan. New Delhi: Pearson, 2007. 521–29. Print.
"Vandana Shiva Joins Dharna against POSCO Project." *The Hindu*. 23 June 2011. Web. 13 June 2013.

NOTES

1. The paper is not intended to delegitimize the promise of Third World environmentalism, nor does it seek to trivialize questions of life and livelihood on which most of these movements are based. Rather it attempts to see the contradictions and unevenness in the theorizing and delivery of these movements. Unlike conventional notions of Third World environmentalism which is seen as a political commitment against capitalist development, the present paper offers a more problematic politicization of such movements, not just in their challenge to global capitalism, but also within those movements themselves and their competing notions of concepts like "development" and "people."

2. The Greek term *zoe* refers the simple fact of life common to humans and animals. It is distinguished from *bios* which is the way of living proper to an human individual. *Zoe* was never given importance by philosophers like Aristotle who found man's essence in his politics. With its simple fact of living, *zoe* is more about reproduction (like animals) and has no role to play in social and political lives. I have borrowed this idea from Giorgio Agamben (1998).

3. The situation in Niyamagiri vis-à-vis Vedanta Aluminium Limited (VAL), which created conditions for the thesis of the present paper, has radically changed. The Supreme Court of India had earlier ordered a referendum to know the wishes of the people (consisting of Dongria Kandhas and other *adivasis*) around Niyamagiri Hills and find out if mining will infringe upon the religious and cultural rights of the tribals living there. Since then all the twelve *gram sabhas* (village councils) have rejected mining activity in Niyamagiri. After this setback, VAL is understood to have applied for mining rights in other hills like Kurlapeta, Chandalgiri, and others, which are in the same region and close to Niyamagiri.

SIX

From Bhopal to Biometrics[1]

Biological Citizenship in the Age of Globalization

Pramod K. Nayar

This essay is interested in the corporeality of the "Global South" and its insertion into a global ecology of tissues, organs and gene lines. If natural resources from countries in the "Global South" are ruthlessly exploited within the circuits of global industrial-military capital, then one of the tasks of ecocriticism ought to be the critique of "tissue economies," as Waldby and Mitchell term the imbrication of biological matter with global capital. This component of ecocriticism which the present essay seeks to develop has to do with the very politics of survival. Bodies and corporeality in the Global South, suggests the essay, are engaged in the politics of survival because their bodies are differentially valued resources in the global marketplace of organs and tissues. By focusing on the materiality of bodies and the construction of ontologies within global discourses—whether of citizenship, compensation, cell lines, or biodiversity—the essay calls for a return to materialist ecocriticism. The ecology of tissues and organs, stem cells and genetic matter within circuits of capital determine the material realities of bodies in the Global South. Material ecocriticism, in Serenella Iovino's terms, "takes matter as a *text*, as a site of narrativity, a storied matter, a corporeal palimpsest in which stories are inscribed" (451, emphasis added). My essay focuses on the matter—or text—closest to us, the body, and examines the *interlinked* narratives, stories, discourses *and* material contexts—toxicity, material aid, surgery—that construct the meaning of matter. I extend, as should be clear, Iovino's materialist ecocriticism to the material bodies of humans. Materialist ecocriticism, this essay assumes, needs to move beyond the textual and look at other "matter/s," such as the human body, whose meanings, like in the case of texts, are generated through their interaction with and situatedness in specific material contexts. This essay is therefore interested in bodily matter/s, and the fraught terrain of ecocriticism that it maps (even minimally given the scope of the essay) includes as its topographical features: global prospecting, banking, marketing of the body and its constituent matter (from tissues to DNA), the differential (racialized) evaluation of injured bodies, biometric identification where the body returns as the irreducible sign/text and repository of human identity. Together these become the material substrate of a biological citizenship enmeshed within global circuits of research, exchange and capital.

Capitalism in the age of globalization has invested heavily in biologicals—genes, tissues and organs. This essay argues that biological citizenship in the new age of capital results in human bodies and organs being increasingly enmeshed in the circulation of global finance, and organs are capitalized upon in the age of tissue economies, often with tragic consequences for bodies from countries like India. My thesis is that bodies in the "Global South" lack biological citizenship in their supposedly sovereign countries, because their bodies are inserted into global systems and circuits of exchange. The essay, after a brief discussion of biological citizenship, traces the emergence of the new biological citizenship in India back to the Bhopal gas tragedy. It then examines transplantation, organ trade—mainly in the black market—tissue and stem cell banking and biocommons. Bioprospecting and surveillance constitute the practices and discourses that now define our very ontologies. If, having "investigated" and found that the corporeal "matter" of Bhopalis isn't quite in need of compensation, if the material "body" of ecoterrorist Warren Anderson (to whom I shall return later in the essay) remains secured away from justice-seekers and if particular cell-lines are of potential interest to certain research laboratories then a scrutiny of these practices approximates to a material ecocritical practice where bodies are turned into texts. In the final section it turns to biometrics and the databasing of individuals in surveillance and informational societies.

A. INTRODUCING BIOLOGICAL CITIZENSHIP

Discussions of citizenship are invariably centred around its national form and location. Citizenship relies on specific "biological presuppositions, [that] explicitly or implicitly, have underlain many citizenship projects, shaped conceptions of what it means to be a citizen, and underpinned distinctions between actual, potential, troublesome and impossible citizens" (Rose and Novas). Contemporary biological citizenship does not take a racialized and nationalized form alone. As bioprospecting and biopiracy show, not all have equal citizenship in this new biological age. Anthropologist Adriana Petryna's work among Chernobyl victims demonstrated how biological citizenship embodies a demand for particular protections, for the enactment, cessation or modification of particular policies or actions such as insurance and compensation "based on medical, scientific, and legal criteria that both acknowledge biological injury and compensate for it" (4).

However, in the globalized age, biovalues for many in India today are determined *outside* the nation's geopolitical borders. In the age of the virtual, electronic communications and online lives corporeality and personality, identity and character are increasingly embodied today, but it is an embodiment that locates identity and citizenship *outside* the nation state. Biological citizenship here is a manifestation of the power of global capital and the erosion of national sovereignty as bodies get inscribed into shadowy networks of organ trade, bioprospecting, bio-commerce and even biocommons. Recalling the fungible bodies of slaves and labor from the older forms of Empire, the new biological citizenship inscribes "national" bodies into global exploitative networks. In other words, the test of a nation's sovereignty is not only with regard to its agency over economic or defence policy but lies in the ascription of biovalue to its national citizens, from victims of industrial disaster to organ trade to biometrics.

B. BHOPAL, THE POLITICS OF COMPENSATION, AND BIOVALUE

In order for us to understand biovalue and biological citizenship in the age of globalization, let us take a look at compensation awarded in two cases: India's Bhopal and the USA's 9/11.

In December 1984, a plume of toxic gas leaked from an American-owned Union Carbide pesticide plant in Bhopal. The leak killed 2,000 people almost instantly, and thousands more died later from the aftereffects of the toxic gas, an ingredient in pesticides the plant produced. By official reckoning, 578,000 people were affected. That was the number finally awarded compensation by specially created claims tribunals. There were no criminal convictions stemming from the leak until 7 June 2010, when eight former executives of the company's Indian subsidiary were convicted of negligence. The men were sentenced two years in prison and fined 100,000 rupees, or $2,100. According to figures published by the Indian government's apex medical research organization, the Indian Council of Medical Research (ICMR), at least 20,000 people had died by 2009 from injuries caused by the gas leak. ICMR's research shows also that between 1984 and 1989 there were 3,500 spontaneous abortions as a result of gas exposure. These need to be taken into account. In the Curative Petition filed in the Supreme Court, the Indian government has ignored the findings of its own research agency and presented a ridiculously low figure of 5,295 deaths caused by the gas disaster. This does not include the inherited impairments and health problems. (Documents on the tragedy, including investigative reports, may be found at http://bhopal.net/source-documents/, accessed 14 Aug. 2013)

The 9/11 attacks left 2977 victims dead. The compensation paid was $2,082,128. It left 6300 victims and the compensation paid ranged from $500 to $8.6 million. The statistics are from the U.S. government website "September 11 Victim Compensation Fund 2001" (http://www.justice.gov/archive/victimcompensation/payments_deceased.html). In sharp contrast is Bhopal. There are 22,900 dead since 1984, and the compensation paid was $1,917. For the 542,210 injured the compensation was $494. Over 27 years of suffering it comes to Rs 2.50 or 5¢ per day. Each American life was valued at more than a thousand times higher than a Bhopali life. The distinction points to, also, the financial politics of bodily survival in different parts of the world and the juridical-legal interventions—by way of compensation—into these politics.

The politics of compensation alerts us to questions of biovalue, of how life is differently valued across the world, and how death is measured differently across the world, as Jacques Derrida puts it in his response to 9/11 (2003). But this question of biovalue is not really about differential compensation alone. It is about the denial of biological citizenship, despite proven and proveable bodily injuries, to Indian victims of the Bhopal disaster.

As a result of the flows of global capital we institute new biovalues that have nothing to do with national citizenship rights or claims. The Bhopal victims' bodies across generations have not accrued biovalue within India because their biovalues are determined outside it, embedded in global circuits of finance, capital and labor. This becomes the new instantiation of globalization where bodies and their biovalues, previously essential to and seamlessly articulated within a national frame, have suddenly been altered by global forces, with little or no state intervention in their biological citizenship. Victims have been globalized, even as globalization has refused them rights and claims *as* victims. Bhopalis, in effect, have suffered a biomedical but also civil death in that the Indian state has not supported their biological citizenship. The denial or erosion of the biovalue of injury effects a major shift in the perception of bodies and therefore of citizenship itself. The Indian citizens of Bhopal have been reterritorialized in terms of bodies and injuries because their biovalues have been recalibrated (only) in terms of the *liability* they present to the transnational company.[2]

Here we see a different dimension of biological citizenship emerging. First, the materiality of local bodies is highlighted when they contribute to the economics of globalization—as labor. It is native bodies that construct the very structures and processes—the embodiment, so to speak, of the global industrial-chemical-financial complex. Second, this materiality of local bodies in the case of Bhopal is split along two lines: a) their injuries, maiming, and deaths foregrounded their materiality as victim-bodies; and b) the processes of compensation and treatment dematerializes their bodies and the extent of injuries. What we see emerging in the new cultures of biological citizenship is a political citizenship without a simultaneous entry into any national biological citizenship—because this last is located outside the national borders.

The campaign for justice for Bhopal victims has recognized and made use of this increased awareness of biological citizenship.

The posters tell the story of ecocide and biological citizenship in powerful ways. Pablo Bartholomew's award-winning photograph of the dead child being buried shows a dead body (available at http://www.netphotograph.com/pablo/bhopal/, accessed 14 Aug. 2013). But ecocide, as we know, isn't only about the dead: it is about survivors as well, as in these posters suggest. One "Bhopal Marathon" poster (available at http://www.bhopalmarathon.org/cryforbhopal/files/12.html, accessed 14 Aug. 2013) shows a boy with horribly deformed legs with a cricket bat. The name "Sachin" of course invokes India's, and arguably cricketing history's, greatest batsman ever, Sachin Tendulkar. The "marathon," with echoes of some of the world's best-known marathons (New York, Boston, Mumbai), is savagely ironic here: the boy in the poster with his impaired body will *not* be running in any of them. The write-up on the side gives another sense of the marathon: in the case of the Bhopal victims, their damaged bodies measure the marathon not in the miles they will never run but in the years they have spent fighting to be even *recognized* as impaired bodies. The idea of the "Bhopal Marathon" captures in one simple idiomatic expression and poster the "slow violence" Rob Nixon has identified as the core feature of all ecocides. Nixon uses the term to describe the "formless threats whose fatal repercussions are dispersed across space and time . . ." where "toxic buildup, massing greenhouse gases, or desertification may be cataclysmic, but they're scientifically convoluted cataclysms in which casualties are deferred, often for generations" ("Slow Violence" 14). For the Bhopalis with impaired bodies, it has indeed been a marathon of pain, suffering, poverty, humiliation, and the quest for justice, both in India and transnationally. Their bodies have been trained to accept this slow violence, as the series of poster campaigns titled "Torture Me" suggest. The posters invite the perpetrators:

> Throw acid in my eyes. Play a blowtorch down my throat. Strip the lining from my lungs. Choke my child to death in front of me. Drown me in my own blood . . . (http://www.studentsforbhopal.org/node/55, accessed 14 Aug. 2013)

In addition to demonstrating the absence of biological citizenship for their impaired bodies, the Bhopal campaign also drew attention to another body: that of Warren Anderson, then chairman and CEO of Union Carbide.[3] The survivor bodies locate their problems in the transnational organization, but also in a particular body. Just as the USA embodied all evil in Bin Laden, the Bhopal victims also see their problems as *embodied* in Anderson. Further, as the protestors here and in the USA made clear, Anderson's physical body needs to be relocated to India to stand trial as a murderer.

In the USA, likewise, activist Diane Wilson's campgain to bring Anderson to justice also resulted in posters demanding his "body" as a murdering, ecocidal body (news reports about Wilson's campaigns available at http://www.bhopal.net/oldsite/archive/august2002.html#diane wilson, accessed 14 Aug. 2013). In India the Bhopalis foreground the *need* for the ecocidal body of Warren Anderson (available at http://www.cbc.ca/news/world/story/2009/07/31/bhopal-disaster-arrest-carbide.html, accessed 14 Aug. 2013.). By situating the USA's quest for Osama bin Laden and the Bhopali quest for Warren Anderson the Bhopal protest alongside each other, the Bhopali campaign aligned so-called "Islamic terrorism" (embodied, for the USA, in bin Laden) with *ecoterrorism*, embodied for the Bhopalis in Warren Anderson. Here the posters explicitly signal, I believe, the commonality of terrorism, whether of the religious or the ecological kind.

The awareness of biological citizenship that these posters reflect is, clearly, located with a larger geopolitical terrain where one form of terrorism (Islamic) is deemed completely unacceptable (to the USA/First World) but ecoterrorism targeting the Global South is quietly ignored. If, as Elizabeth DeLoughrey tells it, postcolonial ecocriticism is concerned with the "biotic contours of empire and its discursive inscriptions" ("Quantum Landscapes,"63) then the Bhopal posters show the discursive inscriptions of the Global North upon the passive, unresisting and impaired bodies of the Global South. The Bhopal posters' traumatic bodies, if read as texts, constitute the biotic articulations of the differently hefted definitions of—and here is a short inventory—ecological disaster, compensation, mortal injury, impairment, Western industrial modernity, postcolonial sovereignty.[4] The posters signal, in other words, the fraught terrain of cultural ecocriticism.

Biological citizenship, as is made evident by the posters documenting wrongs and seeking justice for Bhopalis, is citizenship under erasure, yet under constant surveillance. The state's campaign against the coalition for Bhopal victims instantiates the struggle for biological citizenship driven by desperation. Further, the debates over India's sovereignty as a nation also begin to revolve around biological citizenship, notably of the Bhopal victims. After the transnational evaluation of victims' biovalues, the next stage in the new biological citizenship is the organ, gene, and stem cell harvesting of citizenry.

C. BIOCOMMERCE AND THE BIOMEDICAL COMMONS

I now turn to the globalization of bodies and organs in two very different contexts, biocommerce and biocommons. As many as one hundred and fifty parts of the human body are reusable, and according to a 2000 study an individual human body is worth $23,000 on the open market (Hedges and Gaines, cited in Sharp, *Strange Harvest* style 11). From American Type Culture Collection (http://www.atcc.org/), which calls itself a 'non-profit global bioresource centre,' one can now order cell lines, just as one can buy sperm or blood. The body has been informatized for commodification, consumption and circulation within global circuits of commerce. But the commodification scene has a tragic aspect in India.

Every year an estimated 100,000 Indians die waiting for organs, with only 3,500 reported transplant operations. The new legislation on human organ transplants (August 2011) is an attempt to curb commercial transactions in organs. Promising payment or incentives for living donations of organs and tissue is now prohibited. This is based upon the ideal that people should donate their organs and tissues for altruistic reasons, or for love. This is the standard, local, or ground-reality of organ transplants in India.

And yet, India's organs are globalized. Take the case of Amit Kumar, a.k.a Kidney King. Kumar and his associates removed kidneys from poor patients and sold them to paying customers. After he was arrested in 2008, it was established that he was part of a network that supplied kidneys to First World buyers. India is at the center of a massive global bioprospecting movement, where the Indian poor are located in a tissue economy that respects no national sovereignty. If the Bhopal story reveals a complete disconnect between the theory of Indian national, geopolitical sovereignty and the offshore and globalized arbitration of biovalue, the organ transplant racket with its shadowy international instantiates a new "empire" of organs. According to a study cited in 2006 (Lesley Sharp, *Strange Harvest*—be more explicit about the author's name and the title of the article) the international tissue market alone was worth $500 million annually in 2000. In India, where the poor body is regularly mined for organs by those like Amit Kumar, along with food security one needs to speak of organ security as well.[5]

Strangely, as Lesley Sharp notes, the organ transplant ideology relies on a denial of transplantation as a form of body commodification (8). And yet, as I have suggested above, the black market trade in organs, a ground reality in India if reports are to be believed, works precisely on the principle that organs are commodities in a global marketplace. The Indian body has been corporatized, and transplants are not determined by compassion, charity, or love. The social value of the donor and donee are therefore recalibrated under such conditions.

But this organ trade, international in scope, and which reinscribes the poor Indian body into a global marketplace of organs, dovetails into something totally distinct: the biodiversity and the biomedical commons movement that also seeks to globalize bodies. I use the term "inscribe" to gesture at the *value coding* of corporeality and its constituents—blood, tissue, organs—and even functions (surrogate wombs, which I have not addressed here).

Hyderabad city recently concluded its UN Convention on Biological Diversity (CBD), in October 2012. The UN Charter on Biodiversity's Article 15 on the access to genetic resources states:

> Recognizing the sovereign rights of States over their natural resources, the authority to determine access to genetic resources rests with the national governments and is subject to national legislation. 2. Each Contracting Party shall endeavour to create conditions to facilitate access to genetic resources for environmentally sound uses by other Contracting Parties and not to impose restrictions that run counter to the objectives of this Convention.

It concludes with:

> Each Contracting Party shall take legislative, administrative or policy measures, as appropriate, and in accordance with Articles 16 and 19 and, where necessary, through the financial mechanism established by Articles 20 and 21 with the aim of sharing in a fair and equitable way the results of research and development and the benefits arising from the commercial and other utilization of genetic resources with the Contracting Party providing such resources. Such sharing shall be upon mutually agreed terms. (http://www.cbd.int/kb/record/article/6897?FreeText=genetic)

This biodiversity initiative is the making of a biomedical commons. The ostensible aim is to resist the commodification of biological material like tissue and genetic material, arguing against patents on genes and gene sequences that are seen as leading to the enslavement of human bodies (*Tissue Economies* 143). The aim of open source biomedical database is ostensibly also to resist the incorporation of biovalue into globalized and standardized production, distribution and consumption patterns (145).

Open source initiatives—headed by organizations like Open Bioinformatic Foundation, the Bioinformatics Organization and the SNP Consortium allow free web access to DNA, RNA, protein sequences and provide software (like BioPerl) for "reading" genetic material. The Tropical Diseases Initiative is also open source, driven by the idea that to *share* research is to ensure some measure of therapeutic discoveries, embodied in the 2000 joint statement on human genome research by President Clinton and Prime Minister Tony Blair where they claimed that "unencumbered access to this information [human genomic databases] will promote discoveries that will reduce the burden of disease, improve health around the world, and enhance the quality of life for all mankind" (Waldby and Mitchell, *Tissue Economies*, 136).

Such biobanks and biomedical commons constitute a transnationalization of life itself. Biobanks are not merely the objects of governance, they are something through which the *governance of life* is effected (Gottweis, "Biobanks in Action"), but a governance that operates across species and also, crucially for my argument, across national borders. Worldwide diversity biobanks such as those proposed by CBD are not about genomic medicine alone. They institute a new order of biosurveillance where data about individuals, ethnic groups and communities are collected—often without informed consent as we now know—and rendered into disseminable or saleable data. As such, biopolitics has now gone deeper into bodies, to the level of molecular structure and information.

The monitoring of genetic materials and their eventual organization into databanks, even in the guise of medical benefit for all, constitutes a new biological citizenship as well where particular groups have their data ordered and analyzed outside the nation-state's purview the moment the state signs the biodiversity agreement. It entails the management of serum, genes, proteins, blood and other bodily data by institutes and corporations outside the nation-state. This means, essentially, the body has been dematerialized into numbers for absorption into a global matrix of such biodata. Hence my thesis is that the decorporealization of bodies into data constitutes the politics of the *dispersed* body whose data lies somewhere in the vaults outside the nation state under the guise of human diversity.

Bioprospecting for organs and for genetic medicine and research are both preliminary moments to biosurveillance. Biometrics embodies the technology of such a surveillance.

I would like to conclude this section by returning to the idea of "inscription" made above. The discursive apparatuses—biomedical, corporations, legal—that pronounce certain bodies, body parts and body functions as valuable determine our very ontologies. "Inscription," therefore, I take as the construction of our ontologies within such discourses of biomedicine that claims, or proposes, that matter *matters* in particular ways. I am explicitly referencing here the science studies of Karen Barad who incalling for a shift from representationalism to performativity proposes that we link performativity "not only to the formation of the subject but also to the production of the matter of bodies" ("Posthumanist Performativity" 808). Barad refers to laboratories, policies (such as the "diversity" document above), laws (the court ruling in the case of Moore) as apparatuses where they "are not mere static arrangements *in* the world, but rather *apparatuses are dynamic (re)configurings of the world, specific agential practices/intra-actions/performances through which specific exclusionary boundaries are enacted*" (816, emphasis in original). Barad's welcome return to materiality is crucial for an understanding of inscriptions that exclude, include, buy, sell, hire and dismember bodies. For my purposes, therefore, documents such as the ones I have cited are material practices and not simply "texts," and any worthwhile postcolonial *materialist* ecocritical reading must account for these texts given their key role in assigning differential values—what Barad refers to as "exclusionary boundaries"—to bodies globally. Bioprospecting is made possible through, and in turn makes possible, such corporeal inscriptions.

D. BIOMETRICS AND GLOBAL BIOLOGICAL CITIZENSHIP

The cell line, the genetic code, and the tissue signature are only deeper extensions of the global emphasis on biosurveillance that originates in fingerprinting and has now reached the stage of iris prints, voice signatures, and other biometric methods of monitoring and surveillance.

DNA collection is now a common practice in several countries. Biometrics with the databasing of DNA fingerprinting or iris prints is linked less to the larger debate on individual corporeal privacy than to the global interest in borders of nations, bodies and institutions. My biometric data at the U.S. Consulate is now linked to a global database that the immigration official at a U.S. port of entry can access at a click to validate my claim about who I say I am. All bodies are now part of a transnational biological citizenship with biometrics.

Biometric identification is often taken as unique to individuals alone—iris and voice scans, DNA fingerprinting, etc. More importantly these are *mobile* features that we cannot ever, try as we might, leave behind, forget, or erase. Thus we carry our unique identities with us, willingly or unwillingly, wherever we go. Biometrics, as I have argued in my 2012 article "I Sing the Body Biometric: Identity, Identification, Surveillance and Biological Citizenship," can be seen as a response to the increased "dematerialization" of the human in the age of the virtual: the body returns as the foundation of the human. Corporeal integrity is ensured at least in the form of the consistent dataset that is prepared and which only uses those biologicals that do not alter, remains coherent and integrated even as the rest of the body alters with age and environment. Thus, on the one hand, face-to-face interactions and socialization shift to the virtual, with avatars and profile-management online, biometrics reinstates the material body within these same electronic networks in the form of corporeal data.

The body is the biometric border, a "portable border par excellence, carried by mobile bodies at the very same time as it is deployed to divide bodies at international boundaries, airports, railway stations, on subways or city streets, in the office or the neighbourhood" (Amoore, "Biometric Borders" 338). Databases sort people into risk categories and "low risk" individuals fast-tracked while "high risk" ones are searched and even detained. Mobilities are now linked to the body's data, and this primarily affects migrant workers and potential immigrants. Foreign nationals in the UK are subject to a separate biometric identity card scheme named "Biometric Residence Permits" (BRPs) including international students, visiting scholars, entrepreneurs, investors and domestic workers. Migrants, as studies show, saw the BRPs as a constant reminder of their difference, and this significantly affected their everyday lives in the receiving society, and this affected their sense of belonging in the global age because their bodies are documented as *different* (Warren and Mavroudi, "Managing Surveillance?"). This implies that the organization of welfare and social benefits hinge upon the "voluntary" sharing of corporeal and other data with the state or any agency empowered by the state. In the globalized world so much continues to hinge not on mobilities but on monitored, regulated and "databased" corporeal mobilities.

This means, the technique of biosurveillance is a mode of governance itself, and is therefore treated by some commentators as a new form of law making where national/international and private/public distinctions are rendered irrelevant in reconstituting the power relations between individuals, individuals and organizations and amongst organizations (Backer, "Global Panopticism"). The state asks private agencies to share data (the frequent requests to Google to share its data is a case in point), thus blurring the state/market boundaries. The Indian government uses this data, acquired by private agencies, to consider new regulatory laws. In other words, without a formal lawmaking process the state is able to engineer regulation, and even overcome the limits of public lawmaking.[6]

What is important is to note how biological citizenship is the consequence not of a centralized artifact that records our faces/irises/thumbprints but the dispersal of surveillance into our everyday. Ambient intelligence and surveillance technologies ensure that we now live in the age of distributed surveillance. ATMs, locks, cameras, and keyboards add to the biometric state and the surveillance society where your identity is distributed and validated an infinite number of times every hour. It is the *convergence* of multiple databases—from the online shopping you do to the iris scan, from Consulate offices to Google—that constitutes an individual's identity disseminated across borders today. Biological citizenship thus is not simply iris scan and your face, it is the sum total of several such bits of data that comprise you. Data collected by one source is often compared and collated with other sources in order to build a larger, clearer picture of what you are and what you do. Informal and local surveillance are increasingly embedded in surveillance networks that are global and not always under the control of the state or local actors.

When such metadata about an individual goes into massive databases such as the ones Google has, then we have a biological citizenship far beyond national borders. Google, for example, makes its money by using our profiles—which includes personal details, the shopping choices we make, dating, medical searches, travel plans—to present us with advertisements keyed to the words we search and therefore supposedly keyed to our personalities, desires, and requirements. As Siva Vaidhyanathan summarizes it:

> [google] collects information from us when we use its services; it copies and makes available trivial or harmful information about us that lies in disparate corners of the Internet; and it actively captures images of public spaces around the world, opening potentially embarrassing or private scenes to scrutiny by strangers—or, sometimes worse, by loved ones. (84)

Our choices and curiosities that make us what we are, are now in databases somewhere, to be retrieved at will by people with access to the database. Data converges upon matter, the body, even as the body is disseminated as zeroes and ones into the database.

What we see then is the slow absorption of the biologicals into global circuits of exchange and commerce. Biological suffering (Bhopal) is evaluated across the border, even halfway across the globe. Organs and genetic material are traded in licit and illicit ways with no respect for national borders or sovereignty. State intervention, or its absence, in organ transplants, biobanks, clinical regimes, databases of its populations, and of course its responses—ethical, financial, but also social in the form of support structures, welfare and employment—to suffering render us citizens of a new biological citizenship. Genetic data is acquired and stored elsewhere, outside national borders. Biometric data is collated, along with vast amounts of other information about personalities, character-traits and individual behaviors, into databases dispersed across the world. So, in the new cultures of biological citizenship our bodies are broken down into data to be collated in an elsewhere that has nothing to do with nations and national identity. From the construction of biovalue by MNCs in Bhopal to the compilation and sorting of biologicals in global software and surveillance culture, we have come a considerable way. In terms of "national story-telling," then, Bhopal, organ trade and biometrics have a rather tragic story to tell where our bodies are not our own any more. The "organized irresponsibility" that Ulrich Beck identified in his 1998 study as central to contemporary risk societies (18)—and he is speaking specifically about environmental degradation and how nobody is really held accountable for what happens—is demonstrated in the case of the new cultures of biological citizenship where no one, not even the state of which these are political citizens, owns up for "denatured" bodies.

We have just been corporeally globalized.

A material ecocritical practice starts with matter. And this makes it imperative that we start with the matter closest to us: our bodies. The Bhopal campaign that encodes a social memory of the abuse of matter—memory here is taken, after contemporary memory studies (Radstone and Hodgkin, *Regimes of Memory*), as material with the symbols of ruin and decay circulating across cultural pathways of production and consumption, as in the posters—the documents about biodiversity that invest matter with particular values and the databasing of populations that exclude or include matter constitute, this essay has proposed, a fit subject for material ecocritical work. Examining the apparatuses within which differential, uneven, and oppressive biological citizenship is forged, along with their ontologies, becomes, in the age of tissue economies, biodiversity and biometrics, a political project against the slow violence of creeping biosurveillance.

WORKS CITED

Amoore, Louise. "Biometric Borders: Governing Mobilities in the War on Terror." *Political Geography* 25 (2006): 336–51. Print.

Andrews, Lori, and Dorothy Nelkin. *Body Bazaar: The Market for Human Tissue in the Biotechnology Age*. New York: Crown, 2001. Print.

Backer, L.C. "Global Panopticism: States, Corporations, and the Governance Effects of Monitoring Regimes." *Indiana Journal of Global Legal Studies* 15.1 (2008): 101–48. Print.

Barad, Karen. "Posthumanist Performativity: Toward an Understanding of How Matter Comes to Matter." *Signs* 28.3 (2003): 801–31. Print.

Beck, Ulrich. "Politics of Risk Society." *The Politics of Risk Society*. Ed. Jane Franklin. Cambridge, UK: *Polity*, 1998. 9–22. Print.

DeLoughrey, Elizabeth. "Quantum Landscapes: A 'Ventriloquism of Spirit.'" *Interventions* 9.1 (2007): 62–82. Print.

Derrida, Jacques. "Autoimmunity: Real and Symbolic Suicides." *Philosophy in a Time of Terror: Dialogues with Jürgen Habermas and Jacques Derrida*. Ed. G. Borradori. Chicago and London: U of Chicago P, 2003. Print.

Gottweis, Herbert. "Biobanks in Action: New Strategies in the Governance of Life." *Biobanks: Governance in Comparative Perspectives*. Ed. Herbert Gottweis and Alan Petersen. New York: Routledge, 2008. 22–38. Print.

Iovino, Serenella. "Stories from the Thick of Things: Introducing Material Ecocriticism." *Interdisciplinary Studies in Literature and Environment* 19.3 (Summer 2012): 449–60. Print.

Mukherjee, Suroopa. *Surviving Bhopal: Dancing Bodies, Written Texts, and Oral Testimonials of Women in the Wake of an Industrial Disaster*. London: Palgrave-Macmillan, 2010. Print.

Nayar, Pramod K. "I Sing the Body Biometric: Identity, Identification, Surveillance and Biological Citizenship." *EPW* 96.32 (2012): 17–22. Print.

Nixon, Rob. "Slow Violence, Gender, and the Environmentalism of the Poor." *Journal of Commonwealth and Postcolonial Studies* 13.2 and 14.1 (2006–07): 14-37. Print.

Petryna, Adriana. *Life Exposed: Biological Citizenship after Chernobyl*. Princeton, NJ: Princeton UP, 2002. Print.

Radstone, Sussanah, and Katherine Hodgkin, eds. *Regimes of Memory*. London and New York: Routledge, 2003. Print.

Rose, Nikolas, and Carlos Novas. "Biological Citizenship."*Global Assemblages: Technology, Politics, and Ethics as Anthropological Problems*. Ed. Aihwa Ong and Stephen J. Collier. Malden, MA: Blackwell, 2005. 439–64. Print.

Sharp, Lesley. *Strange Harvest: Organ Transplant, Denatured Bodies, and the Transformed Self*. Berkeley and London: U of California P, 2006. Print.

Vaidhyanathan, Shiva. *The Googlization of Everything (and Why We Should Worry)*. Berkeley and London: U of California P, 2011. Print.

Waldby, Catherine, and R. Mitchell. *Tissue Economies: Blood, Organs, and Cell Lines in Late Capitalism*. Durham and London: Duke UP, 2006. Print.

Warren, Adam, and Elizabeth Mavroudi. "Managing Surveillance? The Impact of Biometric Residence Permits on UK Migrants." *Journal of Ethnic and Migration Studies* 37.9 (2011): 1495–1511. Print.

Women Seek Nod to Sell Kidneys." *The Hindu* 13 November 2012: 6. Print.

NOTES

1. An earlier version of this essay was delivered at the Sahitya Akademi seminar on "Indian Civilization and its Discontents: Literature, Culture, and Society" under the title "Biosurveillance, Biobanks, Biocapitalism and their Discontents: The New Cultures of Biological Citizenship," Puducherry, India, 23–24 November 2012. I am grateful to Sudhir Kakkar, Anup Dhar and Nandana Dutta for responses to the paper. Thanks also to Swarnalatha and Scott Slovic for suggestions, and to Scott for thoughtfully sharing the special issue of *ISLE* on material ecocriticism.

2. It is therefore no surprise that the Bhopal survivors have expressed their support for the Kundankulam struggle against nuclear power plants. The removal of the nuclear liability clause from India's recent treaty with the USA again sacrifices the claims of native biological citizenship at the altar of global capital.

3. Anderson left India under cover of darkness, literally, after the disaster. Demands to bring him back to India to stand trial have never produced any results, and Anderson continues to live peaceably in the USA, as the Bhopal victims have pointed out.

4. In her study of testimonials from the Bhopal tragedy Suroopa Mukherjee documents the use of dancing—and bodies-in-performance—by the survivors as a means of memorialization and resistance (see chapter 7).

5. As I write this, I read a newspaper account where the wives of six convicts have requested the Andhra Pradesh State Human Rights Commission for permission to sell their kidneys to raise money for the release of their husbands (*The Hindu* 13 November 2012: 6).

6. This kind of transnational surveillance has another component to it. The behaviour of states and MNCs have also been subject to scrutiny through information-gathering. Take the non-governmental organization, Transparency International, for example. It describes itself as "a global network including more than 90 locally established national chapters and chapters . . . Politically non-partisan, TI does not undertake investigations of alleged corruption or expose individual cases, but at times will work in coalition with organisations that do" (Transparency International, *About Us*, http://www.transparency.org/about_us).

SEVEN

Beyond the Eco-flaneur's Footsteps

Perambulatory Narration in Zakes Mda's Ways of Dying

Laura A. White

Speaking about his understanding of place and how place functions in his writing, Zakes Mda explains that when he writes novels, he usually begins with place, which he considers a living, dynamic aspect of his work: "The place is important in my fiction because it becomes part of the story. It's not just there as background. Sometimes the place functions as character" ("An interview" 69). Literary critics have been attentive to Mda's use of space in his first novel *Ways of Dying* (1995), which is set in an un-named South African port city in the transition period at the end of the apartheid era. They have investigated Mda's representation of the city, developing innovative ways to think about the creative potential of human movement through urban spaces; however, they have largely neglected Mda's understanding of place as an animate element of his fiction and have failed to consider the city as a living environment constituted through human and nonhuman interaction.[1] Conversely, this novel with an urban setting has not drawn much attention from ecocritically oriented scholars who have tended to focus on Mda's later novels, leaving Mda's representation of the city as environment under-explored.[2]

Rather than suggesting that *Ways of Dying* has little to contribute to environmental discussions, I would argue that the critical response to this novel evidences the need for an expanded ecocritical practice that reflects the environmental priorities of the Global South. As the new field of postcolonial ecocriticism continues to develop, scholars are revising definitions of nature and environment to reflect that urban poverty, waste management, housing and urban development are some of the most crucial environmental issues facing the Global South. These scholars are also demanding attention to ways that these issues are imbricated in a colonial history and a global modernity that materially and ideologically influence human relationships with the land.[3] Building on this work, I propose that in *Ways of Dying* Mda uses the perambulations of his main character, Toloki, to articulate the environmental interests of black South Africans living in urban slums. Mda's narration of Toloki's physical and temporal movement presents the city not as inert backdrop but as vital assemblage, constituted by and shared among present and past human and nonhuman inhabitants. This strategy enables Mda to illustrate the contemporary environmental priorities of the urban poor in relation to a history of black South Africans' alienation from the physical land and from mainstream South African environmentalism. As he choreographs Toloki's movement, Mda delivers powerful commentary about the material treatment of the nonhuman environment, about unequal access to the land and its resources, and about South Africans' abilities to imagine a self that is immersed in co-constitutive relationships with the nonhuman environment.

A. THEORIZING WALKING

Toloki is a mobile observer of the city who serves to make hidden histories, vitalities and interdependencies narratable. Theorists and historians of walking from Michel de Certeau to Rebecca Solnit have observed that walking and narrating share common territory, and the language of walking preserves traces of this association. For instance, perambulation, commonly used as a synonym for walking, has a history that the Oxford English Dictionary traces back to 1425. At that time it designated not simply pedestrian travel, but "the action or ceremony of officially walking round a territory (as a forest, manor, parish, etc.) to determine and record its boundaries, to preserve rights of possession, etc., or to confer a blessing" (1a). Historically, then, perambulation united the ambling human body with processes of verbal and visual recording, specifically the composition of narrative descriptions and pictorial maps that served to establish the borders of forests and identify the territory to which laws applied. A novel involves a type of perambulation as a writer verbally paces out a territory to bring it under control, and as he sets Toloki in motion, Mda uses the walks of his focal character to disrupt previously recorded boundaries, re-distribute the authority to compose the city, and re-define "rights of possession" for the urban poor. Reading Toloki's travels in this way, against a Western history of walking, also illuminates the need to address barriers that have kept the walking and storytelling practices of the Global South absent from many discussions of the relationship between walking and narrative. As Solnit acknowledges in her impressive study that catalogs walkers from pilgrims to protestors to prostitutes, although her book carries the subtitle, "A History of Walking," it primarily reflects the experiences of post-industrial, First-World wanderers who choose to walk (267). On its own, such work is not adequate for the challenge of deciphering Toloki's journeys, but bringing research about urban South Africa into conversation with European and American theorizations of urban wanderers/narrators can clarify how Mda deploys Toloki's treks to communicate the environmental concerns of black South African residents of urban slums. Attending to the pedestrians of the Global South might make broader contributions to scholarship about walking as well; following the paths and practices of these walkers could complicate assumptions about home and private property, community membership and narrative authority that persist within dominant theories and quietly perpetuate social and environmental injustice.

Scholarship devoted to African cities provides a useful starting point for investigating relationships between South Africa's urban spaces and the human inhabitants who move within them. For instance, Abdou Maliq Simone develops a concept of infrastructure that addresses human interaction with the city; as opposed to the typical understanding of infrastructure as a physical network of roads, Simone re-orients the term to describe "the ability of residents to engage complex combinations of objects, spaces, persons and practices" to form conjunctions (qtd. in Nuttall 39). Investigating literary representations of urban South Africa in a study titled *Entanglements* (2009), Sarah Nuttall borrows Simone's concept and divides the urban literary spaces she studies into four infrastructures: the street, the café, the suburb and the campus, and she allies these spaces with a distinctive figure that circulates through each: the migrant worker, the aging white man, the illegal immigrant, and the hustler. As opposed to the single figure of the flaneur, this model recognizes that different identities emerge from complex engagements between objects, spaces, persons and practices, but it does not account for nonhuman life or allow for consideration of the city as living environment; further, although it does treat the South African city as more than background, the tactic of dividing the spaces and figures that populate them for separate discussion seems to reinforce their isolation rather than their entanglement as inter-dependent components of the same city.

Nuttall emphasizes that analyzing the South African city with this model of infrastructures addresses aspects of urban experience that are inaccessible to the traditional flaneur. Citing Amin and Thrift, she argues that "invocations of the flaneur in urban theory can underestimate the extent to which striating openness and flow are a whole series of rules, conventions and institutions of regulation and control" (qtd. in Nuttall 38). In the context of post-apartheid South Africa, Nuttall reminds us that these rules and institutions have included legal restrictions and formal and informal practices of surveillance that continue to influence ways that inhabitants relate to urban spaces (39). She is right to demand that urban scholarship attend to historical and contemporary contexts that influence how individuals conceptualize and utilize the space of the South African city. However, I suggest that unlike the flaneur, the eco-flaneur can allow for this kind of contextualized study. In fact, the figure of the eco-flaneur might serve to remedy some of the blindspots inherent in the traditional flaneur's urban wanderings while also enabling more nuanced attention to the dynamics of human interdependencies with the nonhuman world, therefore addressing a gap that lingers in Nuttall's valuable formulation of urban entanglement.

In *Writing for an Endangered World* (2001), Lawrence Buell elaborates his concept of the eco-flaneur in reference to an Anglo-American literary tradition. His concept involves more than adding the prefix eco- to conventional notions of the flaneur, and it might be productively adapted to a South African context in ways that address concerns that Nuttall raises about the limitations of the flaneur. Traditional definitions identify the flaneur as "a man who takes visual possession of the city" (Wilson, qtd. in Buell 65). However, formulating his concept of the eco-flaneur, Buell suggests that understanding the flaneur as a man who encounters the city as a detached consciousness, recording experiences of "fascinated alienation," reflects only one particular strand of urban writing; he contends that a counter-tradition can be identified, one in which writers endow their wanderers with "something like an environmental consciousness" (89). Buell distinguishes this observer from other narrators of the metropolis: the eco-flaneur represents the city in eco-systemic terms and recognizes the city as habitat. Further and most significant for my purposes, narrating the city as habitat changes the understanding of the city as background space for human life and forces a re-imagination of the borders of personal identity, as the eco-flaneur emerges from "perpetually guarded sanctuary into a state of vulnerable, porous transpersonal reciprocity with people and place" (89). This encounter that Buell describes as urban transpersonality challenges understandings of physical boundaries between bodies, and also refigures temporal boundaries. Referencing Georg Simmel's claim that "a person does not end with the limits of his physical body or with the area to which his physical activity is immediately confined but embraces, rather, the totality of meaningful effects which emanate from him temporally and spatially" (106), Buell argues that urban transpersonality demands "participation in community, both with fellow inhabitants in the present and with past generations" (84). Therefore, like Nuttall's attention to entanglements and Simone's theorization of infrastructure, Buell's formulation of the eco-flaneur focuses attention on shifting, complex interactions in ways that demand re-thinking static notions of identity. Unlike these other models though, the eco-flaneur's experience of reciprocity between people and place and his experience of the city as habitat encourages greater attention to long term interdependencies between the human and nonhuman lives that mutually constitute the city.

B.TREADING THE GROUND OF HISTORY

A South African eco-flaneur, then, is not a detached walker, but a figure whose spatial and temporal movement through the city makes an alternative perspective on capitalist modernity—including long-term interdependencies and inequalities—narratable. As opposed to other figures that may wander the city—the exile, the migrant, the glaneur[4]—the eco-flaneur represents not only movement, but the potential to narratively construct the city; the steps of the eco-flaneur are an experiential articulation of the temporal, spatial and material entanglements of the city. Elaborating her own conception of an eco-flaneur who can function as a "marginal-critical figure" in the lineage of the flaneur, Catriona Sandilands urges attention to historical specificity (37). She argues that the flaneur's "mode of critical reflection emerged from specific spatial and perceptual relations . . . his was a specific position in the historic city and crowd" (37), and formulating a concept of a South African eco-flaneur similarly demands careful attention to how a method of narrating the city emerges from the "specific spatial and perceptual relations" experienced in post-apartheid South African cities.

The South African eco-flaneur walks through a city whose spatial and perceptual possibilities have been shaped by a history of European colonialism. The colonial settlement of South Africa began as a garden; instead of a permanent settlement, the Dutch East India Company envisioned it as a food provisioning station. As Allister Sparks explains in his study, *The Mind of South Africa* (2006), Jan van Riebeeck, celebrated by Afrikaners as the founder of the colony, needed to find a way to keep his community separate from the land and the people of Africa. His supervisors suggested building a canal, which would literally make the white inhabitants their own island; lacking the manpower for such a project, Van Riebeeck instead decided to plant a hedge of almond trees that would surround his community and accomplish the same goal of isolating them from the rest of Africa (Sparks xvi). Sparks uses the hedge as an analogy for the mindset of South Africans, laying the foundation for his own project of exploring the thinking of those on either side of the hedge. However, for my project, the materiality of the hedge and the way it initiates a pattern of marking boundaries holds significance. The early settlers decided to create a garden on African soil, which necessitated the creation of a boundary wall. To mark the border of their garden, they used a living wall, this hedge of almonds that remains growing in the

Kirstenbosch Botanic Gardens today, three hundred and fifty years later. They claimed the land and its bounty for themselves, visibly etched their ownership and concept of private property into the land, and exiled native Africans from their garden, beginning an alienation of black South Africans from the land that, like the hedge, continues to this day.

This colonial logic of erecting borders and separating populations extends into contemporary urban planning: it was institutionalized by apartheid policies that have been discontinued, yet still exert influence on patterns of land ownership and thinking about the land. A.J. Christopher's analysis of the spatial dimensions of apartheid in *The Atlas of Changing South Africa* (2000) connects early colonial classification to later apartheid policies. Christopher explains that the National Party swept to victory in the 1948 elections, campaigning on the promise to extend and codify earlier colonial practices of racial segregation. In 1950, the Population Registration Act "provided for the compulsory classification of the population into discrete and racially defined groups . . . enforced by statute and specified on identity documents" (101). So the classificatory thinking that allowed flora and fauna to be named and cataloged was extended to humans, who were forced to carry identity cards and assigned places to live based on racial classification. Urban planners were employed to create separate spaces for each group; Christopher verifies that by 1970 residential segregation was nearing completion in most cities, and by 1980 it was virtually total (128). Apartheid policies of classification and separation have also left an imprint on environmental discourse and practice in South Africa. As McDonald explains, "thousands of black South Africans were forcibly removed from their ancestral lands to make way for game parks, and billions of rands were spent on preserving wildlife and protecting wild flowers while people in 'townships' and 'homelands' lived without adequate food, shelter, and clean water" (1). Such practices led to a perception of environmentalism as a white, upper/middle class suburban issue, blind to or even hostile to urban, black South Africans, and they reinforced a discourse in which black South Africans, when they were acknowledged, were figured as threat to the environment (McDonald 1; Khan 17–24). Writing in the 1990s as apartheid policies were being dismantled and new spatial and human relationships were being imagined, Mda's narration of the city and its gardens involves more than just his vivid descriptions of flowers; it reflects the power of naming and organizing and the way that these legacies of colonialism and apartheid continue to influence South Africans' relationships with the land.

B. THROUGH THE EYES OF THE ECO-FLANEUR

Turning to Mda's novel, the figure of the eco-flaneur creates a useful framework for thinking about Toloki's physical and imaginative travels. As Toloki strolls through the city, deserted for the Christmas holiday,

> he struts like a king, for today the whole city belongs to him. He owns the wide tarmac roads, the skyscrapers, the traffic lights, the flowers on the sidewalks. That is what he loves most about this city. It is a garden city with flowers and well-tended shrubs and bushes growing at every conceivable place. In all seasons, blossoms fill the air. Sometimes when he goes to a funeral he picks a flower or two as long as no one sees him. (Mda 46)

This moment has drawn much attention from literary critics, who often focus on the first two lines of the passage to remark on Toloki's proprietary attitude toward the city as he struts like a king through an area of the city once denied to him. Theorizations of the eco-flaneur can productively complicate this understanding of Toloki's relationship with the city by enabling an investigation of how Toloki constructs himself through transpersonal reciprocity with a city that encompasses various past and present inhabitants. As he moves through the city, Toloki does not wander with the gaze of a detached observer; he participates in a sensory and emotional encounter whose ramifications extend beyond the present moment. In the opening lines of this passage, Toloki does seem to exult in a feeling of ownership, but in the following lines that feeling dissolves into an expression of affinity with the flowers. He acknowledges joy and awe at his visual and olfactory contact with the flowers, and he registers the flowers' active participation in constructing the city as they grow, bloom and "fill the air." As the narrator elaborates on Toloki's experience of moving through the city, Toloki's desire to possess the city becomes deflated by his surreptitious plucking of the flowers; his current strutting posture is cast in relation to past restrictions which Toloki still carries with him as he walks. Unlike the historical flaneur who moved in a crowd, Toloki wanders through a deserted city, and in part, his current feeling of belonging depends on the lack of other human inhabitants to contest his claim. While the city is currently devoid of human residents, this moment is also populated with Toloki's memories of times when he moves through the city with far less confidence, approaching the flowers "sometimes," only when he feels no one is watching him. Toloki's past subjection to the surveilling eye positions him as a walker who is both observer and observed; within this singular moment of walking, he displays his participation in shifting and contradictory relationships with the city's past and present, its human and nonhuman inhabitants.

Toloki's journey from the business district, filled with flowers at all seasons, to the informal settlement on the city's fringe demonstrates not only that the city is a living environment, but also that it is an environment where resources and risks are not distributed equally. In the informal settlement, Toloki walks among shacks made from cardboard, plastic, and corrugated iron. While blossoming flowers fill the city with color and fragrance, in the settlement, the colors derive from the multi-colored plastics woven into the shacks and the scent comes from "the quagmire of dirty water and human ordure that runs through the streets" (48). Serving as the inverse of the bountiful garden, the informal settlement is characterized by waste: instead of flowers, it showcases discarded plastic and metal, polluted water and barren ground. Mda's fictional depiction of this anti-garden echoes Mamphela Ramphele's academic study of the settlements: "The squalor and sterility of the black segments of the urban areas is shocking. The only "flower species" is the ubiquitous multi-colored plastic litter, blown around by the wind from heaps of garbage rotting in the streets" (5). Descriptions that relate and conflate colorful garbage and flowers emerge repeatedly in the novel, with Toloki's friend Noria holding flowers he has brought her from the city as she stands in the settlement amidst piles of debris and with the shack that Noria and Toloki build with discarded materials depicted with brilliant colors that shimmer in the sun like flowers (Mda 50, 67, 107). These descriptions destabilize divisions between nature and culture and complicate fixed conceptions of flowers by narrating various, contextualized interactions among humans, objects and nonhuman nature. They also call attention to the distribution of environmental assets and environmental risks. Contrary to the fascinated alienation of a flaneur or the isolated figures of Nuttall's infrastructures, the movement of the eco-flaneur circulates embodied encounters that span multiple sections of the city, demanding attention to ways that the bounty and beauty of the downtown garden and the sterility of the settlement are intertwined.

Additionally, as Mda repeats the word transport throughout the novel to refer to Toloki's flights of imagination as well as his travel in taxis and on trains, Mda insists on connections between material and discursive relationships to the land's resources. Toloki's alienation from the gardens of the city involves not only limited material access to the land, but also constraints on his ability to imagine himself as participating in long term reciprocity with the land. Toloki uses imaginative walks to confront this dual alienation and to re-configure restrictive notions of the garden and the human. For example, when he helps his friend Noria to rebuild her shack, Toloki brings copies of *Home and Garden* magazine and plasters the inside of the shack with pictures of gardens. Then, he shows Noria how they can walk in the garden they have created. In their minds, they leave her makeshift shack, cobbled from scraps and enter the garden, "a potpourri of color, designed by expert landscape architects" and filled with beds of snapdragons, hollyhocks, salvia, azaleas, and petunias (113). The fantasy walk gives them power, allowing them to transform the concept of the garden as an exclusive, classificatory space by layering interactions between different material spaces and their discursive representations. These representations also span time, ranging from the magazine's marketing and commodification of the garden in contemporary South Africa through to the garden's colonial history and van Riebeek's cultivation of a garden hedge to mark productive land from "waste land." Toloki and Noria inhabit their garden and then blend the imagined garden into the life of the settlement as Noria remarks that the extra pages of the magazine "will come in handy . . . when she sits on the floor since her shack is devoid of even a single stool" (114). Rather than escaping into a fantasy world of luxury, Toloki and Noria use imaginative transport to redefine the garden and to claim their right to share the aesthetic, material and discursive resources that the city offers. Responding to a legacy of alienation, Toloki and Noria craft their own version of the garden and shape their own image of themselves as active participants in long term, reciprocal relationships with the urban environment. Bringing such encounters into focus, theorizations of the eco-flaneur's reciprocity with the environment can enhance understandings of the city that do not register its living dynamics or account for long term material and discursive interactions. Other models can make cities a focus rather than a backdrop, but the eco-flaneur can reflect cities as living subjects with agency, history, and ecological limits and can provoke reflection on how humans imagine themselves in relation to these.

C. REDEFINING INHABITANCE

Toloki's physical and temporal movement composes a map of the city from his "marginal-critical position," uncovering an alternate mode of inhabitance in the lives of Noria and Toloki, but also implicating established practices of land ownership in environmental injustices. Flashbacks narrate Toloki's encounters with Nefolovhodwe, who left the village a poor man and made his fortune in the city by selling coffins. Toloki visits Nefolovhodwe's home and finds it "surrounded by a tall security fence, which had warnings that it was electrified attached. There was a well trimmed hedge inside the fence" (127). Like van Riebeek, Nefolovhodwe has built a hedge; he updates the simple enclosure with a modern touch, an electrified fence to surround his garden and protect his wealth and his house full of extravagant possessions, even from the friends who helped him when he was poor. Mda's narration follows Toloki from Nefolovhodwe's enclosed space to the impoverished settlement in order to demonstrate that simply changing the color of the man behind the hedge leaves the logic of exclusion unchallenged. In the flashback, Toloki leaves the house of Nefolovhodwe thinking that he has lost all respect for this man (166). Mda uses movement through time to connect spaces, so that as Toloki exits this memory, he enters the dumping ground in the present. In contrast to the protected garden space filled with beauty and luxury, the dumping ground is a stark space, a two room settlement shack surrounded by mud. The barren land and the name "the dumping ground" characterize the space and the people who occupy it as devalued waste; however, this place also challenges the label of waste by illustrating a different set of values than Nefolovhodwe's garden. Although visually stark, the dumping ground is an open space where Madimbhaza, Noria, and other women of the settlement share their resources to feed, clothe and care for children who have been orphaned or abandoned by their parents. Toloki realizes that inhabiting this place is a choice the women have made; they could be working in the city for wages but have chosen instead to share their time and resources to provide for the children.

The narration of these areas as dynamic environments rather than abstract, background spaces conveys a powerful critique of the idea of land as private property to be isolated and exploited for the good of an individual owner. Nefolovhodwe is ridiculed by the narrator and rejected by Toloki; although like Toloki, he is a poor migrant from the country, the fact that he arrives in the city and embraces capitalism and individualism and secures his property, even from his former friends, demonstrates that a perpetuation of dominant attitudes toward land as private resource constitutes a problem. While the land is in black rather than white hands, the underlying attitude of individual profit remains the same. However, Toloki recognizes a different potential in the organization of the women in the settlement. Although they have been alienated from productive land, they are not helpless victims. They have agency which they exercise by making the deliberate choice to reject capitalist imperatives of wage labor, consumption, and individualism in favor of shared community resources. By moving directly from the home of Nefolovhodwe to the dumping ground, Mda's narration forces an awareness that these two places exist in relation with one another, a responsibility that the garden hedge attempts to obscure. Instead of a simple idealization of community values that valorizes the virtues of the poor, Mda demands recognition that poverty and waste are produced by the logic of capitalism and private property.

Writing his novel as South Africa was preparing for a democratic future, Mda uses perambulatory narration to re-map the city and to caution readers that transforming South Africans' relationships with urban environments requires re-imagining cities in terms of interdependencies. Instead of policies that isolate urban housing or waste management as the problems of the poor, Mda urges a comprehensive rethinking of human relation with the nonhuman environment that takes into account the imbrication of wealth in poverty, the past in the present, and the ideological in material configurations of the city. More than a decade since the novel's publication, as South Africans mark the centenary of the Land Act of 1913—legislation that restricted black land ownership—frustrations about the failure of post-apartheid land reform policies abound. As contemporary scholars study global efforts to alleviate urban poverty, Mike Davis warns that most solutions are seriously flawed: valorizing the creativity of slum dwellers can be used as an excuse to abandon them to their own problems and negate the state's responsibility to address poverty and housing, while governments and NGOs may champion market-driven solutions that actually perpetuate inequalities (72, 82). Mda's novel does not provide simple remedies to these complex problems, but it predicts the failure of models that refuse to fundamentally re-think inhabitance, and it offers the footsteps of the eco-flaneur as one point of entry for this challenging task.

WORKS CITED

Barnard, Rita. *Apartheid and Beyond: South African Writers and the Politics of Place*. New York: Oxford UP, 2007. Print.

———. "On Laughter, The Grotesque, and the South African Transition: Zakes Mda's *Ways of Dying*." *Novel* 37.3 (2004): 277–302. Print.

Buell, Lawrence. *Writing for an Endangered World: Literature, Culture and Environment in the US and Beyond*. Cambridge, MA: Harvard UP, 2001. Print.

Caminero-Santangelo, Byron. "In Place: Tourism, Cosmopolitan Bioregionalism, and Zakes Mda's *The Heart of Redness*." *Postcolonial Ecologies: Literatures of the Environment*. Ed. Elizabeth DeLoughery and George B. Handley. New York: Oxford UP, 2011. 291–307. Print.

Christopher, A.J. *The Atlas of Changing South Africa*. New York: Routledge, 2000. Print.

Davis, Mike. *Planet of Slums*. New York: Verso, 2006. Print.

Feldbrugge, Astrid. "The Human and the Non-human World in Zakes Mda's *The Heart of Redness* and *The Whale Caller*." *Local Natures, Global Responsibilities*. Ed. Nancy Grimm, Ines Detmers, and Katrin Thomson. Amsterdam: Rodopi, 2010. 151–66. Print.

Khan, Fareeda. "The Roots of Environmental Racism and the Rise of Environmental Justice in the 1990s." *Environmental Justice in South Africa*. Ed. David A. McDonald. Athens, OH: Ohio UP, 2002. 15–48. Print.

McDonald, David A. "Introduction: What is Environmental Justice?" *Environmental Justice in South Africa*. Ed. David A. McDonald. Athens, OH: Ohio UP, 2002. 1–12. Print.

Mda, Zakes. "An Interview with Elly Williams." *Missouri Review* 28.2 (2005): 62–79. Print.

———. *Ways of Dying*. New York: Picador, 1995. Print.

Nixon, Rob. "Environmentalism and Postcolonialism." *Postcolonial Studies and Beyond*. Ed. Ania Loomba, Suvir Kaul, Matti Bunzl, Antoinette Burton, and Jed Esty. Durham, NC: Duke UP, 2005. 233–51. Print.

———. *Slow Violence and the Environmentalism of the Poor*. Cambridge, MA: Harvard UP, 2011. Print.

Nuttall, Sarah. *Entanglement: Literary and Cultural Reflections on Post-Apartheid*. Johannesburg: Wits UP, 2009. Print.

Ramphele, Mamphela. *Restoring the Land: Environment and Change in Post-Apartheid South Africa*. London: Panos Publications, 1992. Print.

Robinson, Jennifer. "(Im)mobilizing Space: Dreaming of Change." *Blank: Architecture, Apartheid, and After*. Ed. Hilton Judin and Ivan Vladislavic. Cape Town: David Philip, 1999. Print.

Sandilands, Catriona. "Green Things in the Garbage: Ecocritical Gleaning in Walter Benjamin's Arcades." *Ecocritical Theory: New European Approaches*. Ed. Axel Goodbody and Kate Rigby. Charlottesville: U of Virginia P, 2011. 30–42. Print.

Sewlall, Harry. "Ecofeminism in Zakes Mda's *The Heart of Redness*." *New Directions in Ecofeminist Literary Criticism*. Ed. Andrea Campbell. Newcastle upon Tyne: Cambridge Scholars Press, 2008. 108–23. Print.

Steinwand, Jonathan. "What the Whales Would Tell Us: Cetacean Communication in Novels By Witi Ihimaera, Linda Hogan, Zakes Mda and Amitav Ghosh." *Postcolonial Ecologies: Literatures of the Environment*. Ed. Elizabeth DeLoughery and George B. Handley. New York: Oxford UP, 2011. 182–99. Print.

Solnit, Rebecca. *Wanderlust: A History of Walking*. New York: Penguin, 2000. Print.

Sparks, Allister. *The Mind of South Africa: The Story of the Rise and Fall of Apartheid*. Johannesburg: Jonathan Ball, 2006. Print.

Vital, Anthony. "Situating Ecology in Recent South African Fiction: J.M. Coetzee's *The Lives of Animals* and Zakes Mda's *The Heart of Redness*." *Journal of Southern African Studies* 31.2 (2005): 297–313. Print.

———. "Waste and Postcolonial History: An Ecocritical Reading of J.M. Coetzee's *Age of Iron*." *Environment at the Margins: Literary and Environmental Studies in Africa*. Ed. Byron Caminero-Santangelo and Garth Myers. Athens, OH: Ohio UP, 2011. 185–212. Print.

Wenzel, Marita. "Appropriating Spaces and Transcending Boundaries in *The Africa House* by Christina Lamb and *Ways of Dying* by Zakes Mda." *Journal of Literary Studies* 19.3–4 (2003): 316–30. Print.

Wright, Laura. *Wilderness Into Civilized Shapes: Reading the Postcolonial Environment*. Athens: U of Georgia P, 2010. Print.

NOTES

1. Literary analyses of Mda's *Ways of Dying* that focus on issues of place while failing to fully investigate the dynamics of human interaction with nonhuman nature include Rita Barnard's "On Laughter, the Grotesque and the South African Transition" and *Apartheid and Beyond: South African Writers and the Politics of Place*, Jennifer Robinson's "(Im)Mobilizing Space," Marita Wenzel's "Appropriating Space and Transcending Boundaries in *The Africa House* by Christina Lamb and *Ways of Dying* by Zakes Mda" as well as Sarah Nuttall's *Entanglement*.

2. Ecocriticial analyses of Mda's works tend to center around his novels *The Heart of Redness* and *The Whale Caller*; these include: Vital, "Situating Ecology in Recent South African Fiction: JM Coetzee's *The Lives of Animals* and Zakes Mda's *The Heart of Redness*"; Wright's analysis of *The Heart of Redness* in her book *Wilderness into Civilized Shapes: Reading the Postcolonial Environment*; Sewlall's "Ecofeminism in Zakes Mda's *The Heart of Redness*"; Feldbrugge's "The Human and the Nonhuman World in Zakes Mda's *The Heart of Redness* and *The Whale Caller*"; Steinwand's analysis of *The Whale Caller* in "What the Whales Would Tell Us"; and Caminero-Santangelo's "In Place: Tourism, Cosmopolitan Bioregionalism and Zakes Mda's *The Heart of Redness*."

3. Rob Nixon's essay "Environmentalism and Postcolonialism" represents an early theorization of intersections between ecocriticism and interests of the Global South. He expands on such connections in his book *Slow Violence and the Environmentalism of the Poor*. Particularly significant for my interest in South African environments is the work of Anthony Vital, especially his essays "Situating Ecology in Recent South African Fiction" and "Waste and Postcolonial History: An Ecocritical Reading of J.M. Coetzee's *Age of Iron*" where he urges attention to urban environments and the discourses of South African environmental justice.

4. The glaneur or gleaner collects left-overs from harvested fields and is a figure with an extensive social, legal, and artistic history in the West. From Jean-Francois Millet's 1857 painting *Les Glanueses* (*The Gleaners)* to Agnes Varda's film, *Les Glaneurs et La Glaneuse* (*The Gleaners and I)* released in 2000, artists have documented ways that rural and urban inhabitants glean in various ways, hunting for food scraps to survive or retrieving discarded objects to create art. Toloki certainly shares aspects of the glaneur's ability to creatively re-purpose what others have rejected, but he is more than a gatherer who collects objects and turns them to new uses; his walks turn the city itself to unexpected uses, and the lens of the eco-flaneur helps to make this challenge to dominant ideologies and trajectories legible in a way that other models of walking or gleaning do not fully address.

EIGHT

Reconsidering the Eco-Imperatives of Ukrainian Consciousness

An Introduction to Ukrainian Environmental Literature

Inna Sukhenko

Ukraine is currently undergoing great changes as it moves toward becoming a democratic society, which calls for transformations in all spheres of society. These changes are occurring in a way that reflect the spiritual and intellectual traditions of Ukrainian society. While Ukrainian culture has become more democratic in recent decades, there have been corresponding processes of industrial development and ecological destruction. Like other parts of Eastern Europe (and regions of the Global South in more distant parts of the world), Ukraine is struggling to balance cultural change, economic development, and environmental protection—at present, the latter is faltering. Ukrainian literature, however, is playing a role in directing public consciousness toward the priority of ecological issues. This paper seeks to analyze the ambivalent interactions that occur between people and the nonhuman environment at various levels. It tries to highlight the ironic combination of traditional environmental appreciation and contemporary industrial degradation that have occurred in Ukraine. The paper not only introduces us to a rich body of Ukrainian literature, but also tries to focus on the "ambiguity" that occurs between a renewed ecological consciousness and a continuing ecological degradation. Ukrainian society is currently striving to combine democratization with the persistent memory of post-Soviet approaches to solving social problems. One aspect of this transitional process is the society's effort to overcome typical social indifference on the part of Ukrainian citizens and cultivate a more active sense of civic responsibility and engagement among the public. Ukraine is physically located at the very heart of Europe, mediating between Western and Central Europe to the West and Russia to the North and East. But Ukraine has its own unique cultural traditions and perspectives regarding the natural world, rooted in pre-Christian beliefs and rituals. Traditionally, Ukrainian cultures have been deeply attached to the natural world, but immigration from other parts of Europe and Asia over the course of history have led to significant changes in Ukrainian society, including more detached and destructively materialist attitudes toward nature.

The study of ecological consciousness of Ukraine as a former Soviet country with its changing social processes and systems demonstrates the great concern about ecological problems nowadays. In contrast to more highly developed countries, Ukraine is characterized by a lack of trust in its government's activities. Furthermore, the general lack of ecological consciousness has slowed the rise of social activity in the sphere of environmental protection. The main reason for this situation is the deficient level of public awareness about environmental problems and the people's feeling that they have minimal power to act in protection of the Ukrainian environment. Exposure to information about how more highly developed countries have had positive experiences in adopting environmental protection laws, waste-free production, and environment-friendly technologies would surely have an encouraging influence on public involvement in Ukraine.

Describing the relationship between nature and Ukrainian culture requires distinguishing between the two major parts of Ukraine's bilateral character. This includes the physical environment and its ethnic traditions—this can be called "the eco-ethnic complex." Eco-ethnic factors have given rise to the conflict between industrialization in Ukraine's territories and historically and ethnically rooted nature-oriented spirituality. The tensions between industrialization and nature adoration are the essence of the contemporary environmental predicament in Ukraine (Yermolayeva 63). Ethnic-nature relations impact the specific ethnic character that is regarded as a set of social and psychological features (feelings, values, psychological intentions, emotions, etc.) that are determined by economic, historical as well as geographical factors concerning how society functions and can appear in culture, traditions, and customs.

The distinct features of Ukrainian ethnic character include individualism, introversion, tolerance, and emotional abundance, the complex combination of which influences the formation of Ukrainian eco-consciousness as well as the further development of the country's economy and industrial activity (Yermolayeva 63).

The contemporary ecological situation in Ukraine represents the dynamics of ecological culture formation of the Ukrainian society. Ukrainian ecological culture certainly includes substantial knowledge about nature, as is evident in the agricultural system; the reliance on the seasonal calendar even in urban settings in the twenty-first century; the impact of nature on folk art, folklore, and folk medicine. These are only a few examples of how various kinds of ecological knowledge systems continue to permeate Ukrainian culture, even as the society confronts the challenges of industrialization and economic development.

Ukrainian culture has always been greatly influenced by its natural environment. Its folk wisdom has preserved and transmitted a careful attitude toward nature, based on a practical, rational understanding of how best to use natural resources. The Ukrainian people have not traditionally resisted nature; on the contrary they have always had an emotional attachment to the natural world and a respectful attitude toward nature, according it an equal or sometimes even a superior status. Despite the Ukrainian worldview's changes in recent decades, Ukrainian culture is characterized by its tight, unbreakable links with nature throughout history (Taglina 42).

Ukraine's ecological culture has shifted its trajectory of development from an agricultural focus to a technological focus. This development parallels a transition from the nature-centered beliefs of Slavic heathens through the theologically centered principles of Christianity and anthropocentric attitudes toward nature. This was typical of Ukraine when it was a part of the totalitarian Soviet state, but now it has turned toward a newly restored, nature-centered ecological ethics. The recent formation of ecological ethics is based on a new awareness of the Earth as a crucial life-support system.

In thinking about the development of Ukrainians' new ecological consciousness, it is necessary to pay special attention to the economic causes of both contemporary ecological crises and environmental policy formation. The post-Soviet rise of consumption is based not on environmental resources but on human needs and wishes.

The ethical aspects of human attitudes toward nature in the Ukrainian cultural tradition determine the moral principles of human-nature relations in the Ukrainian worldview. Ukrainian ancestors comprehended a sophisticated system of environmental values. A fundamental love and adoration toward nature is a central feature of the Ukrainian mentality. Widely shared Slavic religious beliefs emphasized a non-dualistic understanding of the relationships between humans and the non-human world and stressed the equal value of all living creatures. Adopting Christianity and Western European philosophical ideas promoted the sense of humans' dominant role in human-nature relations in what has come to be contemporary Ukraine. Christianity, with its clearly defined anthropocentric perspective, justified dualistic principles of human-nature relations and paved the way for environmental exploitation. Due to the new Christian philosophy, environmental adoration began to be substituted for the more rational, exploitative approach towards nature. However

Christianity, as practiced in Eastern Europe, has also called for responsibility not only in interpersonal relations but also in human relationships toward all natural phenomena (Anreyeva 202).

The ecology-centered tendency of Ukrainian literary studies as a separate trend asserted itself in the 1970s and 1980s under the name "ecological aesthetics" (Boreyko and Pustovit 17). The field was a part of aesthetic studies, devoted to researching relations between humans, nature, and the biosphere which occurred during a period of deep ecological crisis due to the dynamic growth of the country's industrial zones and the destructive use of natural resources, posing an extreme threat to human life as well as cultural memory. This was a period when acute ecological problems typical throughout the territory of the Soviet Union became prevalent in Ukraine. For instance, Government-sponsored projects changed the course of rivers in Siberia, polluted Lake Baikal, and produced human-made seas with the objective of building new hydroelectric plants. The new ecological aesthetics aimed, by contrast, at the development of reasonable behavioral norms, which could regulate human-nature relations in their ecological, cultural, social, and aesthetic aspects in order to keep the balance between environmental and cultural values.

Literature and film played major roles in the process of supporting the evolution of a new ecological culture in Ukraine. For instance, literary critics have identified a recurring motif in Ukrainian fiction that focuses specifically on human-made seas and their destructive consequences and effects. Instances of early literary interventions that offer resistance to ecological catastrophes in the region are the novella *The Poem about the Sea* (1956) by Oleksandr Dovzhenko, the novel *Birds Leaving Their Nests* (1965) by Ivan Chendey and the novella *Farewell to Matyora* (1976) by Valentin Rasputin to name a few.

Oleksandr Dovzhenko's *The Poem about the Sea* was regarded as one of the author's great works is a narrative about "the labor and the beauty" as well as "the ethnos's souls" (Yermolayeva 33). Critics refer to it as proof of the author's "great magnitude of romanticism" (Vargo 96). Later fiction by Oleksandr Dovzhenko as well as works by Ivan Chendey and Valentin Rasputin have been regarded as a fictional treatments of human stories connected with the urgent ecological problems of those times. The most resonant fiction of that period was a novel called *The Cathedral* (1967) by Oles Honchar, where the writer emphasized the long-standing problem of negligent attitudes toward the Samara cathedral—one of the most unique and exquisite cultural memorials in Ukrainian culture. The concern with environmental degradation and negligent treatment of cultural monuments went together and represented a new mid-twentieth-century Ukrainian struggle to restore both cultural phenomena that had been undermined by the Soviet system and the natural environment that had been trashed by Soviet-era industrialization.

Critics (Ukrainian as well as foreign) have begun to pay attention to the ecological features of Ukrainian literature as well as its new problems, themes, and even terminology. The subject of "Human-Nature" has always been and still is among the main problems of Ukrainian literature.

While speaking about the early background of ecological criticism in Ukrainian literary studies, it is necessary to mention the works by Euvgen Malanyuk (1897–1968), Yuriy Lypa (1900–1944), and Dmytro Chyzhevsky (1894–1977) who spoke about the geographical factor as the main "coordinator" of Ukrainian thinking as well as its ecological culture, which is based on Slavic traditions, religion, and morality and which corresponds to the "biological rhythms" of Ukrainian nature. These works did not explicitly mention an "ecological approach" towards narration analysis, but they formed the background for its further development in Ukrainian literary studies. Later Ukrainian fiction analyzed the close human-nature relations. Analysis of fictional works took into account issues of environmental protection in its multilevel representations (including harm of and crime against nature, the preservation of nature for future generations, etc.) which form the ecological culture of a character. Among the first works that talk about an ecological approach toward Ukraininan literature is *The Image of 'Earth' in Canadian and Ukrainian Prose* (1996) by Natalia Ovcharenko, one of the contemporary Ukrainian literary scholars. For Ukrainian literary traditions, "earth" is not only soil; "earth" is life, hope, joy, and wealth according to Natalia Ovcharenko and other scholars of Ukrainian literature.

The ecocritical movement in Ukraine is at its dawning point. The first move toward implementing the ecocritical toolkit for narrative analysis within Ukrainian literary studies was made by Oleksandr Kozlov (1937–2010), one of the brightest theorists of literary criticism in Ukraine. He popularized the ecocriticism movement through his lectures and conference debates. However his illness did not give him the opportunity to publish his articles on ecocriticism. In addition to Natalia Ovcharenko, ecocriticism is among the academic interests of two contemporary Ukrainian literary scholars—Larysa Gorbolis and Mykola Tkachuk. Their writing demonstrates that ecocriticism within the Ukrainian context occurs the crossroads of ecological aesthetics and literary studies, based on religious and social perspectives.

The explosion at the Chernobyl nuclear power plant on April 26, 1986, was a watershed event in the Ukrainian ecological consciousness, but it did not find prompt representation in Ukrainian literature about nature. During the first year, the information available consisted of brief official reports and plenty of gossip about "something has happened" in addition to occasional meetings with people who were evacuated from the Zone. It was only later that this accident was recognized as the most painful and evident symptom of what the Ukrainians came to call the "ill society." This nuclear catastrophe as well as other alarming factors made

Ukrainian society face the necessity of getting informed of the truth about itself and building all spheres of life on the basis of a viable moral structure. The literature of the post-Chernobyl period should be viewed through the lens of how the Chernobyl nuclear accident helps Ukrainian literature address questions of truth and human morality.

The accident at the Chernobyl nuclear power plant produced a peculiar surge of ecologically directed publications and the appearance of "ecological nonfiction/documentaries" (the so-called "writings on ecological affairs") and "ecological memoirs" (representing the "writer's ecological approach"). Ukrainian writers' responses to the nuclear accident were extremely dynamic; among them are documentary notes/nonfiction works such as Yuriy Shcherbak's *Reasons and Consequences* (1986) and *Chernobyl* (1989), Volodymyr Yavorivskiy's *Maria with Mugwort at the End of the Century* (1988), Ivan Druch's *The Chernobyl Madonna* (1988), Borys Oliynyk's *Seven* (1988), Lidiya Viryna's *At That Fire Night* (1989), and Vitaliy Mykulskiy's *Fire Destructors* (1996), among others. But the critical responses to these and other works were restrained and low-key. Sometimes a newly published work went by unnoticed. The reasons for this are well known. In the first years after the Chernobyl tragedy, reporting any information dealing with the accident was prohibited; journalists and literary authors were only allowed to repeat only the official point of view relating to this catastrophe. There was a complete factual blackout regarding the actual event. But some time later, in spite of all obstacles and prohibitions, Ukrainian society became aware of Chernobyl as a technical accident that had global ecological effects, the lingering consequences of which will be experienced by future generations. This awareness of the real events that occurred at Chernobyl reached the mass media partly due to the publication of memoirs of the Chernobyl catastrophe by first-hand witnesses.

All these works—journalistic and literary—differ in form, but they are united through their common subject matter. This diversity of narrative form in writings about Chernobyl gives critics the opportunity to study their common generic features and define "the Chernobyl genre," which was introduced in Ukrainian literature by Marko Pavlyshyn in a work titled *Chernobyl Theme and the Problem of Genre* (1992), which highlights various aspects of the social context of the Chernobyl accident and its aftermath (Yavorivskiy 46).

The stylistic tensions of this "Chernobyl genre" lie in the fact that on the one hand Chernobyl is a real historical event, which many people experienced, but on the other hand this catastrophe is an event which had a much more specific and individualized impact on each author's life. This gives rise to the question about the appropriateness of covering this global catastrophe via a single person's point of view and consciousness. In response to this implicit question, Yuriy Shcherbak created a documentary novel, *Chernobyl* (1989), which contained actual interviews. Volodymyr Yavorivskiy, along similar lines, wrote his novel *Maria with Mugwort at the End of the Century* (1988) by juxtaposing real documents and photos. Representing their authorial perspectives as "honesty" and "humility," these authors state that they did not create fictional texts based on the information about Chernobyl, but instead allowed facts to represent the real situation. Their works sometimes contain the phrase: "After the accident I repeatedly managed to visit the Zone" (Shcherbak 19; Yavorivskiy 16), which can represent a new authorial topos in the literature, a topos emphasizing the author's need to connect time and time again with the historical source of the literary topic. The documentary and multi-perspectival aspect of Ukrainian literary representations of the socio-ecological effects of Chernobyl are highly reminiscent of Japanese author Ishimure Michiko's classic work, *Paradise in the Sea of Sorrow* (1990). By employing a multi-voiced technique merging fiction and journalism, this novel responds to the tragedy of Minamata Disease in the 1960s, another industry-caused contamination which negatively impacted both the natural environment and the human community.

One distinctive aspect of "the Chernobyl genre" deals with documentary literary notes which did not intend to be a journalistic coverage of the catastrophe. "Chernobyl must be a fact of the Ukrainian literature," says Yuriy Shcherbak (19). He framed his series of interviews with people involved in the accident by employing a professional author's "pathetic high style" (Boyko 7).

Even more, Chernobyl can be regarded as a sign of the future within literary and artistic categories, because this contamination event inspired writers and artists from the region to apply their voices and imagination to the meaning of the disaster, just as writers and artists are currently attempting, after nearly thirty years, to develop new aesthetic responses to the phenomenon of global warming. "The Chernobyl epic" was represented as a global tragedy. It needed new literary forms, as Ukrainian scholars have noted: "this form—literary memoirs—appeared and helped to cover the information that was under control by the government" (Dyachenko 211). Literary memoirs not only shaped non-official information about the Chernobyl accident (which was not a single, isolated event but a series of significant disorders and breakages that took place a few days before the main exposure of Reactor 4), but also gave specific details of the social reality of the times. Written by the poets and writers, these literary memoirs depicted the Chernobyl catastrophe and post-Chernobyl events employing a tragic style.

There exists another answer to the question: "Is it appropriate to articulate this global catastrophe from the position of subjectivity?" Since the very beginning of its existence literature has always employed "high" genres and forms in the service of elevated themes. The poets of the Chernobyl period felt that it was the epic genre that best corresponded to the implications of a global catastrophe of such great magnitude. Borys

Oliynyk in his narrative poem "Seven" incorporated the elements of a traditional epic such as myth-making, symbol-making, and hero-creating. Among all other works of "Chernobyl literature," this poem by Boryl Oliynyk was particularly appreciated because of its satire, philosophical background of the problem, and breadth of civil consciousness, which demonstrated close connections between the author and the Chernobyl accident. Borya Oliynik writes:

> This catastrophe turned over the world. The awful scenes are in front of eyes. Duality covered everything. . . . Pessimism transforms into cynicism. . . . The reason is presumptuous human mind and blind faith. . . . The way out is sacrifice and victory of hope over pessimism. The solution is avoidance of duality. Integrity of soul is the basement of existence. (Oliynik 4)

In his poem the poet chooses the form of dialogue between a crow, a representative symbol of cultural pessimism, and an unsuccessful disputant. The subtext of this conversation is Stalin's repression of the progressive-minded people in all spheres of life, who were branded as the "enemies of the Soviet people."

> "You, engineer, a representative of clever minds!" said the crow. "How can you speak about clever minds? Clever minds were shot away as 'enemies of the people' in 1937. If these clever minds were here, we would not have this contaminated air around us and inside us." (Oliynik 4)

This method of narration enabled the author to speak about sensitive issues openly by using non-human characters as mouthpieces.

Ambiguity is another significant aspect of "the Chernobyl genre." While speaking about global relations between humanity and science and about the moral aspect of technological progress, the authors who write within "the Chernobyl genre" do not for the most part describe the detrimental effects of environmental catastrophe on the the planet and the macrocosmos. On the other hand, they depict the harrowing details of chaos and contamination in the microcosmos, using images like a contaminated flower, a cow covered with cellophane, and a mother with her dead child. The reason for this ambiguity regarding scale can be easily explained. The authors who immediately visited the nuclear station after the main exposure within journalists' groups did not realize the size of the catastrophe, and at that moment were confident that the Communist party would control this chaos and put everything in harmonious order. It was only a few years later that they, as well as others concerned, managed to realize the real scale of the catastrophe.

Western scientists considered the Chernobyl catastrophe to be the result of a complex combination of causes ranging from government decisions about using nuclear energy and building the nuclear power plant near Kiev, significant technical changes in building and constructing the nuclear reactor, and lack of responsibility in managing the plant to bureaucratic paralysis and criminal withholding of information about the situation as the first reaction to the catastrophe. All of these factors stemmed from the weaknesses of the whole system in its economic, political, social, and ideological aspects.

"The Chernobyl genre" can also be characterized by the fact that its fact-oriented works are almost entirely devoid of apologists (defending the government and the nuclear authorities). Although the critical vector is directed at the depravity of contemporary society in general, these works do not express a critical attitude toward those who should be responsible for the accident.

The next special feature of "the Chernobyl genre" touches upon the role of literature in this particular post-Chernobyl situation as well as in the contemporary transitional society in general. The authors of that time faced a paradoxical situation: they were a tool of government propaganda, while also understanding that they had an implicit obligation to report the true facts about the Chernobyl accident. In the wake of this unprecedented technological catastrophe, writers realized that they could not adhere to the previous Soviet guidelines for covering public events; their civic responsibility as the voice of the public did not allow them to keep silent. In this awkward position, most writers had to search for a compromise between their two functions; "the writer as a spokesperson of the government" and "the writer as a responsible citizen." This difficult situation sometimes resulted in fictional works where authors publicly apologized and expressed regret for the previous glorification of the industrial projects that led to ecological catastrophes (see, for instance, *The Chernobyl Madonna* by Ivan Druch, which appeared in 1988).

The Chernobyl catastrophe made the whole world quake and was promptly represented in various literary genres: poems and ballads, narratives, stories, and novellas in various languages. After paying a visit to the contamination zone, journalists and literary writers started depicting the tragedy through reportage and essays. They gathered facts and received permission to write about some aspects of the catastrophe. Russian authors were the first to respond to the accident in their works. The 1987 play *The Sarcophagus* by Russian journalist and writer Volodymyr Gubarev is generally thought to be the most famous Russian literary work about Chernobyl. He was among those journalists who arrived at the Chernobyl nuclear station within some hours after the accident. In the preface of his play he wrote:

"In May, immediately after the accident I flew to Chernobyl to write a report. The editor said to me: 'We need honest information. Of any volume. Of any genre.' Unexpectedly the journalistic reportage became a play. A tragedy. But it was not me who chose the genre" (4).This play was staged in more than one hundred theatres in the world and only one in Ukraine (performed by a Russian theatre from Tambov in 1987). In 1989 in the UK this play was honored with the Laurence Olivier Award.

One of the first Russian poems about Chernobyl was the 1988 work called "The Chestnut Silence" by Oleksandr Tkachenko. In depicting the Chernobyl catastrophe, he avoids answering the questions "Who is in charge?" and "What should be the punishment?" but is particularly concerned with how future generations will evaluate Chernobyl. Without being a witness of or participant in the Chernobyl nuclear accident, the author calls for the world audience's attention to the catastrophe.

A similar approach to the tragedy in Ukrainian literature came from Yuriy Shcherbak with his documentary notes and memoir titled *Chernobyl* (1989). As a doctor and a researcher, he reveals his observations, analyzes them, makes certain conclusions, searches for the causes, interrogates those in charge, and suggests an appropriate punishment. Yuriy Shcherbak pronounces a few to be guilty and even reveals their surnames. The paradox of the situation is that, like all other Ukrainian authors who wrote about Chernobyl, Yuriy Shcherbak charges everybody living there who kept silent about the tragedy : "We, who consented to and accepted Chernobyl, are in charge of it; I blame us" (Shchebak 19). Svitlana Yovenko, a Ukrainian poet, supports him: "Carelessness! Inactivity! Sloth! Our ignorance about Chernobyl—all these mean: suicide!" (Yovenko 16).

While stressing the specific threat from numerous nuclear power plants situated in Ukraine and Belarus, Yuriy Shcherbak in *Chernobyl* summarizes the threat to all humanity and offers a final warning: "We came to Chernobyl. We came to a crisis of faith. We came to the edge . . . " (Shcherbak 19).

The brightest representation of the Chernobyl tragedy can be found in the narrative poem "The Chernobyl Madonna" 1988 by Ivan Druch whose son was among those who participated in rescue operations after the accident. Druch's previous set of poems, titled *The Breath of the Nuclear Power Plant* (1978), expressed the author's faith in the peace-devoted use of nuclear power. Although the author lauds the efforts of scientists, engineers, and builders who were involved in the project by casting them in the heroic mode in works like *The Ditch* (1978) and *The Reactor* (1984), he issues an early warning about building a nuclear power plant in the most populated part of Ukraine.

Druch's "The Chernobyl Madonna"was written inhis usual poetic style which combined the artistic and the grotesque: "On the top of the mount there is a nuclear reactor burning, at the bottom of the mount there is a tractor plowing" (44). The audience understood this work as an attempt to represent the Cherynobyl disaster as transformational moment in public attitudes toward technological impacts on environmental and human health. The author emphasizes the specificity of his own position in describing the disaster with the help of a specific poetic toolkit: namely, the topos of humility:, "Aren't you afraid to step on the cosmic road having only your sketchbook in your hand?" (43), the topos of summoning the Muse, "She [the Muse] comes and makes you a poetic pen" (43), and the topos of silence, "I am out of the words" (44). In this poem the author represented his own vision of the tragedy by creating the image of Madonna, God's mother, who knows about the final moments of her son's life but does nothing to prevent him from making his way to the Crucifixion for the sake of humanity. The awareness of the author's personal tragedy as a result of his own son's post-Chernobyl medical condition makes this poem especially poignant.

"The Chernobyl Madonna" has two layers of representation. The first layer is a realistic one, involving the set of realistic portraits like that of the old woman wearing a cellophane dress who secretly walks her cellophane-covered cow in the contamination zone. This episode inspires both humor and fear by highlighting human vulnerability during times of environmental crisis. Having been affected by her contact with the contaminated zone, the old lady tries to kiss a contaminated flower despite being aware of the deadly hazard of the act. Even the risk of contamination cannot prevent her from trying to forge deep connections with the Earth. All of the other characters—a soldier, an atomic scientist, an executive director, a builder, an engineer, and others—are depicted separately, without any ties within the plot, but the readers realize that they all are related to the Reactor and the Zone.

The other layer of representation deals with the poet's lyrical style and philosophical reflections The poet reconsiders how biblical, folkloric, and literary traditions might help bring about a turn in consciousness and help prevent the seemingly unavoidable environmental calamity. The Scythian numens, the usually inconspicuous spirits (manitos) of the ancient Slavs in Ukrainian folklore which are believed to appear under dangerous circumstances to protect the Slavonic people, are portrayed by Ivan Druch as appearing from burial places in order to protect not only the Slavs but the whole of humanity and the Earth from nuclear radiation. His poem is a multilayered mosaic structure which combines the tragic and confessional points of the narration based on real facts and the poet's moral way of reconsidering the catastrophe. The poet concludes with the image of the Madonna holding a baby in her hands: "And this grey-haired Chernobyl Madonna carries this ill planet . . . her own ill child" (Druch 43).

Another eloquent practitioner of the Chernobyl genre is Lina Kostenko, who was among the first visitors to the Chernobyl zone after the contamination. Ten years after the Chernobyl catastrophe, she wrote her novel *The Zone of Estrangement* (1996), where she describes the transformation of the 30-km zone around Chernobyl into a wild, devastated, and deserted zone. In this novel Lina Kostenko raises an important question which the narrative tries to answer: "No, the Chernobyl accident is not the apocalypse, the real apocalypse will come when we forget the Bible and the Holy Word " (Kostenko 5). In her artistic interpretation of biblical images, motifs, and ideas, Kostenko uses God's Word "as a shield and as a weapon in the struggle of good and evil" (Stetsenko 76).

Women writers have been important contributors to "the Chernobyl genre." Of particular interest is the fiction of Svitlana Yovenko. The author depicts "the Chernobyl syndrome" through her female images and characters. The distinctive features of her fiction include female subjectivity, autobiographical and confessional elements, sincerity, directness, fragmentariness, and the female model of narration. Tetyana Kachan, a Ukrainian researcher, comments:

> She [Svitlana Yovenko] outlines a woman's vision of the Chernobyl catastrophe and the role of female writerly activism in the Zone. The image of the Zone, created by Svitlana Yovenko, is dually interpreted—on one hand, it is an area, contaminated with the nuclear radiation, on the other hand, it is a model of the world, restricted the Woman's desires, plans and words. Her Chernobyl is a mosaic of female faces, female actions, female fates. (175)

In her short story "Do Not Scare Me, Baby" (1999), Svitlana Yovenko depicts family problems and the generation gap in the context of the widescale Chernobyl tragedy, again represented through philosophical motives and ethno-historical themes in its multilevel narration. She tries to display the "family war" in the main female character's home against the background of Chernobyl's "invisible" war. Demonstrating the episodes of these "wars"—at the female character's home and in the Zone—the author comes to the conclusion that these wars have a common source of origin: namely, an avariciously anthropocentric human who is in charge of these tragedies. As a result of social and ecological isolation brought about by disintegrating family ties and a spiritually dead society under a repressive government, the female character is ready to die.

In her poem "Explosion" (1987), Svitlana Yovenko stresses the pain and courage of those who dealt with the consequences of the accident in Chernobyl. (Her phrase "the Pain and the Courage" were used as a title of the anthology devoted to the Chernobyl nuclear accident). The poet emphasizes the horrendous situation during the post-catastrophe summer in Kyiv, the capital of Ukraine: "That summer was full of anxiety . . . and Kyiv was free of children" (Yovenko 18). In her stories she emotionally reacts to the jargon of scientists about how "Science requires sacrifices," exhibits concern about the future, and emphasizes the role of collective conscience of the Ukrainian nation in solving the current ecological situation, leading to the reformation of the human soul. What is notable here is how Yovenko specifically instructs Ukrainians to use this ecological catastrophe as motivation to improve themselves and work collectively toward ecological restoration. The poet writes: "Let's look into the sealed wells of the Zone—What can you see there?—You cannot see yourself—Only the words: Stop a fierce and wicked atom!" (Yovenko 18). All of her short stories are permeated with the Chernobyl "signs," which contribute to "the Chernobyl mosaic of faces, lives, fates, actions" (Kachan 175). Her female symbolic vision absorbs biblical reminiscences, touches on Ukrainian folkloric motifs, and expresses both apocalyptic and life-asserting messages. She greatly succeeds in creating the image of "the Zone." In her fiction the Zone is not only the radiated territory but also a model of the Universe where the woman, devoid of human rights, lives. This female vision of the Chernobyl catastrophe and women's place in the Zone enables the unique representation of the Chernobyl syndrome as a psychological condition of distress.

Ecological memoirs devoted to the nuclear accident in Chernobyl are valuable documents which reveal the reasons for the Chernobyl accident to the common people and also enable them to comprehend the consequences of that catastrophe. In unveiling "the Dead Zone," memoir and nonfiction writers confessed that they did not know at that point in time the whole truth about the Chernobyl accident. Even in the twenty-first century, there are works which are subjected to rigorous censorship by the government, such as *Soothsayer from Future* (1996) and *Farewell to Chernobyl* (1991) by M. Malakhuta, a journalist and editor of a regional journal in eastern Ukraine in the Soviet period. A small number of his friends sometimes mention these memoirs, but the author still avoids speaking about them because of negative consequences after the attempt to publish them (Dyachenko 213).

The post-Chernobyl decade's writers tried to focus on the commonly accepted opinions about the accident's consequences, although the ecological problems they depicted allowed the writers to emphasize dangers faced by their human characters. As a result of covering the Chernobyl nuclear accident and its aftermath, most Ukrainian writers of the post-Chernobyl period came to the understanding that every patch of land in the Chernobyl region was dead and inappropriate to live on. Post-Chernobyl nonfiction by Ukrainian writers presents facts and memories reconsidered by the authors and passed through the writers' and eye-witnesses' imagination. However, Ukrainian fiction has not managed to cover the tragedy of the Chernobyl catastrophe to the same extent.

Ukrainian literature—both of the post-Chernobyl period and the contemporary one—is characterized by its ecological-spiritual direction, where ecological and spiritual problems are regarded as an integral whole. This has launched a new paradigm in Ukrainian literary studies that integrates and interrogates the representation of environmental problems in cultural spaces. The new emphasis on representing the moral aspects of the ecological-spiritual problems of Ukrainian society is evidently based on the "cultural memory" theory of Yan Asmann which states that the combination of time and identity shapes the cultural memory of any nation. According to Asmann, a forty-year period is enough for any event to disappear or become a landmark in a nation's memory (Asmann 21). The Chernobyl accident with its post-Chernobyl literature, mainly memoirs, has become a so-called landmark of the Ukrainian ethnic consciousness. All of these factors have created the background for today's dominant tendency in examining human-nature relations in contemporary Ukrainian literature.

The Chernobyl theme in Ukrainian literature is one of the acute sites where the fight for the Truth in literature is taking place in this society. This fight enables Ukrainian literature to demonstrate its civic inclination, its ability to analyze society, and its intellectual honesty. This body of writing about Chernobyl serves to refresh and energize the ethnic and social consciousness of Ukrainian readers and urges citizens to take steps that will avoid "future Chernobyls" by shaping a new cultural consciousness based on strong principles of human morality.

Ethnic peculiarities—uniquely Ukrainian—deal with the dual vision of Chernobyl accident. On one hand, the Chernobyl genre highlights the heroic behavior of firefighters at the nuclear station, the journalists who consciously risked their lives in order to film documentaries about Chernobyl for their descendants, and the civilians who demonstrated their willingness to make sacrifices under such circumstances. On the other hand, the genre reveals the carelessness of building a nuclear station in the densely-populated region near the capital of Ukraine, the mismanagement of its functioning, and the naïve nonchalance of concealing the truth about the causes and the first stages of the catastrophe (Salyga 113). However, perhaps the unique feature of the Chernobyl genre is that the authors tried to reveal the truth of the Chernobyl accident through their poetry and memoirs under circumstances of the total secrecy.

The specific tendencies of the Ukrainian way of thinking should be taken into consideration in order to work out Ukraine's own model of stable economic development. Neglecting these peculiarities will certainly undermine the effectiveness of ecological cultural development and the formation of a new ecological consciousness (Taglina 39).

The analysis of Ukrainian ecological consciousness shows that it is characterized by the lack of trust in governmental and administrative authorities and the low level of public engagement brought about by the lack of information in the environmental protection sphere. Chernobyl literature goads the reader's awareness to focus on the terrifying situation of dead, inert earth in parts of Ukraine's territory (as well as in one of neighboring countries) which makes inhabitation nearly impossible. However, even this dire situation cannot cause the Ukranian people to break their strong ties with the earth since people hold on to the sentiment thatthe earth, though contaminated, is theirs. All Chernobyl literature demonstrates that the archetype of Earth still retains its influential organizing character for Ukrainian consciousness and paradoxically becomes stronger, even under new urban circumstances.

Vargo Oleksandra suggests that in order to overcome the gap between the irrational consumer society and the alternative model of civilization and values, Ukraine should implement ecological behavior and develop ecological culture through discussion of current ecological problems in fiction/nonfiction texts (Vargo 99).Chernobyl literature is closely connected with historical, cultural as well as philosophical contexts in Ukraine. It commemorates those who died immediately in the accident, those who died a slow and painful death after the accident, and those who are dying now—about thirty years after the nuclear catastrophe. Cancer deaths are increasing every year, which comes as no surprise. The inevitability of death becomes a permanent feature of contemporary ecological consciousness of people who live in the contaminated territory (which is much wider than the thirty-kilometer Chernobyl zone).

The process of adopting new eco-imperatives in the society is a multi-staged one. First, scientists dealing with ecological and environmental issues face the necessity of changing the situation. Second, philosophers need to comprehend the complexity of the situation and formulate concepts of stewardship relating to post-Chernobyl human attitudes toward the environment. And finally politicians and non-governmental organizations need to influence the course of events and implement real plans in order to provide not only democratic but also ecological development of the society. According to Anatoliy Tolstoukhov, a Ukrainian politician and scholar, this three-pronged plan of actionwill encourage Ukrainians to implement a new ecological consciousness and forge a new futurefor Ukraine (175).

It is quite early to state that Ukrainian literary studies has the capacity to present well-formed ecocritical research, especially in comparison to the achievements of the North American literary criticism where the content of ecocriticism, its connections with the contemporary ecological thinking, and the methods of implementing the ecocritical toolkit in textual analysis are well developed. Ukrainian literary studies has reached the point of feeling the urgent need to push forward in the direction of ecocritical research, as Ukrainian ecological consciousness, which has achieved a high degree of ethnic self-understanding, is now ready to offer the background for this research ,enabling future Ukrainian artists, writers, and scholars to make an insightful contributions to the field. The history of Ukrainian society, as a Soviet state and a nuclear disaster zone, may seem to defy the possibility of a viable ecologically sensitive society. However, as I have tried to suggest in my article, there is a deep, pre-Soviet tradition of environmental respect and concern among the Ukrainian people, and the excruciating disaster that took place at Chernobyl in 1986 has helped to bring forth the sense of traditional attachment to and concern for nature within this society. Far from being merely crushed—

or irradiated—by this catastrophe, Ukrainians have used Chernobyl to trigger cultural self-awareness and renewed ecological attentiveness. Our literature proves that cultural identity and environmental consciousness go hand in hand.

WORKS CITED

Anreyeva, Tatiana. "The development of moral principles in the aspect of 'human-nature' relations in Ukrainian world perception." *Multiversum. Philosophical Almanac: The Collection of Thesis* 11 (2000): 201–12. Print. [Андреєва, Тетяна. "Розвиток моральних засад відношення «людина-природа» в українському світогляді." *Мультиверсум. Філософський альманах: Збірник наукових праць* 11 (2000): 201–12. Print.]

Asmann, Yan. *Cultural Memory*. Moscow: Slavonic culture languages, 2004. Print. [Асманн, Ян. *Культурная память*. Москва: Языки славянской культуры, 2004. Print.]

Boreyko, Volodymyr and Natalia Pustovit. *Ecological Aesthetics*. Kiev: Logos, 2012. Print. [Борейко, Володимир. Пустовіт, Наталія. *Екологічна естетика*. Київ: Логос, 2012. Print.]

Boyko, Liudmila. "Fictional/Non-fictional epics about Chernobyl: the view on two works." *Soviet Literary Studies* 3 (1988): 3–15. Print. (Бойко Л. Документально-художня епіка Чорнобиля: погляд на два твори // Радянське літературознавство. 3 (1988): 3–15. Print.)

Druch, Ivan. "The Chernobyl Madonna." *Motherland* 1.3 (2011): 5–10. Print. [Драч, Іван. "Чорнобильська Мадонна." *Вітчизна* 1 (1988): 43. Print.]

Dyachenko, Iryna. "Memoir literature about Chernobyl as a way of moral personality's formation." *Science. Religion. Society*. 1 (2008): 210–13. Print. [Дьяченко, Ірина. "Мемуарна література про Чорнобиль як засіб формування моральної особистості." *Наука. Релігія. Суспільство*. 1 (2008): 210–13. Print.]

Gorbolis, Larysa. "Ecocritical dimensions of Ukrainian literature: Expediency and Implementation Eligibility (using the example of Lesya Ukrayinka's "Forest Song")." *Philological treatises* 3 (2011): 5–10. Print. [Горболіс, Лариса. Екокритичні виміри української літератури: доцільність і прийнятність застосування (на прикладі «Лісової пісні» Лесі Українки). *Філологічні трактати*. 3 (2011): 5–10. Print.]

Kachan, Tatiana. "The specificity of tradition and innovation integration in 'female prose' of the 80-ies–90-ies of the XXth century." *Philological Studies. Scholar Journal* 3 (2004): 172–83. Print. [Качак, Тетяна. "Своєрідність інтеграції традицій і новаторства у "жіночій прозі" 80-90-х років XX століття." *Філологічні студії. Науковий часопис* 3 (2004): 172–83. Print.]

Kostenko, Lina. "Estrangement Zone." *Literary Ukraine*. 25 April 1996: 5. Print. [Костенко, Ліна. "Зона відчуження." *Літературна Україна*. 25 квітня 1996: 5. Print.]

Kyselyov, Mykola, et al. *Conceptual Dimensions of Ecological Consciousness*. Kyiv: Paparan, 2003. Print. [Кисельов, Микола, et al. *Концептуальні виміри екологічної свідомості*. Київ: Парапан, 2003. Print.]

Oliynik, Borys. "Seven: Poem." *Literary Ukraine*. 17 September 1987: 4. Print. [Олійник, Борис. "Сім: Поема." *Літературна Україна*. 17 вересня 1987: 4. Print.]

Ovcharenko, Natalya. The Image of 'Earth' in Canadian and Ukrainian Prose. Kyiv: Kyivska Pravda, 1996. Print. [Овчаренко Наталя «Міфологема землі в канадській та українській прозі» Київ: Київська правда, 1996. Print.]

Salyga, Taras. "Poem: the way through the experience." *October* 5 (1988): 110–18. Print. [Салига, Тарас. Поема: шлях крізь досвід. *Жовтень* 5 (1988): 110–18. Print.]

Shchebak, Yuriy. "Chernobyl." *Motherland* 4 (1988): 19. Print. [Щербак, Юрій. "Чорнобиль." *Вітчизна* 4 (1988): 19. Print.]

Stetsenko, Olena. "Bible motives, images and subjects in Lina Kostenko's works". *Bible and Culture* 2 (2000): 73–78. Print. [Стеценко, Олена. "Біблійні мотиви, образи і сюжети у творчості Л. Костенко." *Біблія і культура*. 2. (2000): 73–78. Print.]

Taglina, Yuliya. "National character of Ukrainian ethnos and its impact on the Ecological Culture." *The Herald of Kharkiv National University named after V. Khazarin "Philosophical Peripetia"* 638 (2004): 38–42. Print. [Тагліна, Юлія. "Національний характер українського етносу та його вплив на екологічну культуру." *Вісник Харківського національного університету ім. В. Каразіна «Філософські перипетії»* 638 (2004): 38–42. Print.]

Tkachuk, Mykola. "Human and nature in Ukrainian literature in the focus of ecocriticism." *Dyvoslovo* 6 (2011): 52–56. Print. [Ткачук, Микола. "Людина і природа в українській літературі крізь призму екокритики." *Дивослово* 6 (2011): 52–56. Print.]

Tolstoukhov, Anatoliy. *Globalization. Government. Eco-future*. Kiev: Parapan, 2008. Print. [Толстоухов, Анатолій. *Глобалізація. Влада. Еко - майбутнє*. Київ: Парапан, 2008. Print.]

Tolstoukhov, Anatoliy and Mykola Khilko. *Eco-secure development: The Search of Strategy*. Kiev: Knowledge of Ukraine, 2001. Print. [Толстоухов, Анатолій, Хилько Микола. *Екобезпечний розвиток : пошуки стратегем*. Київ: Знання України, 2001. Print.]

Yavorivskiy, V, ed. *Chernobyl: Tragedy, Gist, Warning*. Moscow: Progress, 1988. Print. [*Чернобыль: трагедия, подвиг, предупреждение*. Под ред. В. Яворовского. Москва: Прогресс, 1988. Print.]

Yermolayeva, A., A.Levtsun and S.Denysenko, eds. *Ukrainian Character: Analytical Survey*. Kiev: Social Studies Centre(Sophia), 2011. Print. [*Украинский характер: аналитический доклад*. Под ред. А. Ермолаева, А. Левцун, С. Денисенко. Киев: Центр социальных исследований (София), (2011). Print.]

Vargo, Oleksandra. "The dynamics of ecological consciousness in Ukraine." *Scholar Herald of Kharkiv State Pedagogical University named after G. Scovoroda* 15 (2004): 95–99. Print. [Варго, Олександра. "Динаміка екологічної свідомості в Україні." *Науковий вісник Харківського державного педагогічного університету ім. Г. Сковороди*. 15. Серія: Філософія. (2004): 95–99. Print.]

Yavorivskiy, Volodymyr. "Maria with Mugwort at the End of the Century." *Motherland* 7 (1987): 16–19. Print. [Яворівський, Володимир. "Марія з полином при кінці століття." *Вітчизна*. 7. (1987): 16–19. Print.]

Yovenko, Svitlana. "Explosion". *Motherland* 5 (1987): 2–21. Print. [Йовенко, Світлана. "Вибух." *Вітчизна* 5 (1987): 2–21. Print.]

NINE

Kissed by Lightning and Fourth Cinema's Naturecultureculture Continuum

Salma Monani

In October 2012 I attended the thirteenth *ImagineNATIVE Film + Media Festival*, hosted annually in Toronto by The Center for Aboriginal Media. As the largest Indigenous film festival in a circuit of approximately sixty-four Indigenous film festivals *ImagineNATIVE* invites Aboriginal filmmakers from across the globe to screen their films, attend workshops, and network with other Indigenous filmmakers (Monani and Brady). The festival premiered with the well-known Abenaki and Canadian artist, Alanis Obamsawin's *The People of the Kattawapiskak River*.[1] In classic Obamsawin style, the film quietly yet unflinchingly documents the housing crisis currently being experienced by the residents of Canada's poorest reserve, Attawapiskat, a community of Cree in Northern Ontario. As I, a citizen of India, followed the camera into overcrowded trailers and leaky homes lacking basic amenities, and listened to the testimonies of residents struggling to find adequate heat and shelter for their families, I could not help but recall the abject conditions of the similarly poor in my own country.

In this article I turn to Fourth Cinema—Indigenous Cinema—as an expression of trans-Indigenous eco-sensibilities that connect the experiences of North American Indigenous people to those in the rest of the Global South. Documentaries such as Obamsawin's *The People of the Kattawapiskak River* speak clearly to the contemporary environmental justice crisis of "where we live, work, and play" faced by Indigenous people across the globe. In this article though, I draw on an example of fictional film, North American Mohawk artist, Shelley Niro's *Kissed by Lightning*, which had screened at the 2009 *ImagineNATIVE* film festival, to suggest additional currents of environmental connections. *Kissed by Lightning* caught my ecocritical attention because it was also included in "Mother Earth in Crisis" themed 2011 Native Film and Video Festival (NAFVF), hosted by the Smithsonian in New York City.

Unlike social documentaries such as Obamsawin's, fictional film often falls into a category of Indigenous film I have previously referred to as obliquely themed (Monani; Monani and Brady). At first glance such films with their characteristic focus on human and individual-centered conflict and resolution seem to have little to do with specific ecological themes, but a close reading of the text, within the extended context of the film-maker's insights reveals otherwise (Monani and Brady). What we see in such films is an ecological consciousness that enmeshes nature into the cultural landscape, not necessarily prioritizing its presence but nonetheless demonstrating its vital connection. Such a fundamental entanglement of material and mental is one that can be understood as a natureculture continuum worth exploring. First, it furthers conversation with the new materialistic turn in ecocriticism begun by scholars such as Stacy Alaimo, Julie Cruikshank, and Joni Adamson.[2] Second, it can also further ecocritical conversations with transnational studies. Specifically, I suggest that such films present what I term are narratives of "alternative eco-contact." The term suggests an alternative to narratives that primarily construe Indigenous eco-sensibilities through explicit, localized struggles for land claims and recalls Paul Lai and Lindsay Claire Smith's transnational concept of "alternative contact" (2010).[3] In doing so, such films broaden the possibilities of trans-Indigenous eco-engagements.

I turn to Niro's *Kissed by Lightning* (hereafter referred to as *KBL*) to illustrate my point because it is definitely not an explicit narrative about land rights. Instead, at its surface level, *KBL* appears myopically focused, following the story of *one* woman's journey through personal loss. However, I argue, that through the interweaving of the highly significant Iroquois legend about 'Hiawatha' and 'The Peacemaker' with Mavis Dogblood's personal life, Niro reveals an eco-sensibility of natureculture that expands her story into the political sphere. In revealing an alternative, oblique narrative of ecological import, *KBL* is one representation of a breadth of Fourth Cinema that has been unduly neglected by ecocinema critics but is nonetheless crucial to understanding contemporary Indigenous eco-expressions. To make sense of this claim, I begin by clarifying what I mean by Fourth Cinema, its trans-Indigenous dimensions, and its ecological possibilities, before engaging with *KBL*.

A. FOURTH CINEMA AS *TRANS*-INDIGENOUS CINEMA

Introducing the term Fourth Cinema in a lecture given in September 2002, Maori filmmaker, Barry Barclay explained:

> I am going to propose here this afternoon that there is a category which can legitimately be called "Fourth Cinema," by which I mean Indigenous Cinema—that's Indigenous with a capital "I" . . .
> . . . The phrase Fourth Cinema comes as a late addition to the First-Second-Third Cinema framework with which you will be familiar, First Cinema being American cinema; Second Cinema Art House cinema; and Third Cinema the cinema of the so-called Third World (9).

While Barclay's 2002 talk and following 2003 *Illusions* article features films that indicate that most Fourth Cinema was emerging from Indigenous filmmakers situated in countries typically recognized as the First World (e.g., *Smoke Signals* by U.S.-based Chris Eyre, *The Fast Runner* by Canada-based Zacharias Kunuk, and *Bedeviled* by Australia-based Tracey Moffat), his categorization is inclusive of *all* Indigenous communities. For example, referring to a tribal protest at a bauxite mine in the Indian state of Orissa, Barclay states: "For the Indigenous people of Orissa State, the cinema camera is out in the surrounding populations, the camera of the First, Second, and Third cinemas. One day, the tribesmen of Orissa State may hold the camera in their own hands. It will not be the camera of the ship's deck" (9).

Barclay's metaphor of Fourth Cinema being held in the hands of Native communities on the "shore" as opposed to the "invaders" on "the ship's deck," or as he describes, those who created the "modern nation state" in which Indigenous people are "ancient remnant cultures persisting" (7) is important. It highlights the potential to reverse a common history of Indigenous peoples across the globe, that is, both the dispossession of their lands and their representation as the "other" by the "invaders'" ethnographic lens.

However, as film scholar Stuart Murray writes, Barclay's project, which forwards a singular concept to unite multiple Indigenous cultures across the globe, "is an inherently complicated process, one plagued by the potential pitfalls and contradictions that accompany any attempt to describe so many different peoples" (11). In fact, Barclay's definition of Fourth Cinema as "outside the national outlook by *definition*" (7)[4] has been scrutinized by scholars such as Houston Wood and Corinn Columpar. In *Unsettling Sights: The Fourth World on Film* (2012), Columpar draws on Wood's *Native Features: Indigenous Films from Around the World* (1995) to point out the somewhat impossible actuality of disentangling Fourth Cinema from practices that imbricate it into a continuum with First, Second, and Third or "invader" Cinema (Columpar xiii).

Such insight cautions against arguments that can limit the possibilities of what Indigenous representation *should* look like. Instead of typing Indigenous films as primarily stories from "remnant cultures persisting" (as Barclay seems to suggest through attention to films like *The Fast Runner*), Columpar argues for complicating representation. As ImagineNATIVE 2012 curator, John Hampton (Chickasaw) states, Indigenous film can challenge "the dichotomy of settler/Indigenous identities" and "reveal that there is more to identity than identifying with one's culture or standing solidly against it." Editors of the recently published *Native Americans on Film: Conversations, Teaching, and Theory* (2013), M. Elise Marubbio and Eric L. Buffalohead, would agree, and suggest that Fourth Cinema accommodates a diversity of possibilities. In "ImagineNATIVE 2012: Ecocinema and the Indigenous Film Festival" (forthcoming), coauthor Miranda Brady and I make a similar case though with particular attention to representations of ecological themes, which span the spectrum from explicit environmental justice documentaries to experimental and fictional films with implied and oblique themes.

While content and genre styles are diverse, most agree that there is one characteristic that collectively marks Fourth Cinema—what scholar Michele Raheja has aptly termed "visual sovereignty." Visual sovereignty highlights strategies by Indigenous people to creatively engage with and reformulate their own representations in culture and through media (Raheja; Marubbio and Buffalohead 8–9). In essence, visual sovereignty speaks to Indigenous agency in self-expression. Ideally such agency is apparent in the whole process of filmmaking, from production to distribution to reception, though many (including Wood and Columpar) point out that, given hegemonic and global networks of filmmaking, this is not always easy to accomplish.[5]

However, even when agency is placed within the context of settler cultures, as Hsingya Huang, Philip Deloria, Laura Furlan and John Gamber point out in their edited special forum of the *Journal of Transnational American Studies* (2012), such expressions of active Indigenous agency can "push against the binary, hierarchical relationship between the major and the minor" and shift the focus from marginalizing Indigeneity to *centering* such experiences (2). Such centering makes room for Indigenous-to-Indigenous relationships as Allen Chadwick writes in the same journal volume and in his subsequent *Trans-Indigenous: Methodologies for Global Native Literary Studies* (2012). As Chadwick explains, if audiences, like the works of the Native artists they engage with, bring Indigenous perspectives to their interpretations, they can forward the centrality of Indigenous knowledge ("A Transnational" 3).

However polyphonic this knowledge might be, engaging with it on its own terms of "seeing, being, and expressing" (Marubbio and Buffalohead 21) is at the heart of trans-Indigenous sensibilities. In essence, the *trans* in "the sense across, beyond, and through" (Chadwick, "A Transnational"; and Chadwick, *Trans-Indigenous* xiv–xv) is a useful tool to illuminate the potential of Fourth cinema as it travels through space and time, and across and beyond its artists' conceptions to Indigenous-centered interpretations amongst Indigenous *and* non-Indigenous audiences alike.

My own work on Indigenous film festivals (Monani; Monani and Brady) explores the concept of trans-Indigenous eco-sensibilities. I argue that such sites, through their collective presentation of multiple Indigenous works from across nations and to public audiences, center diverse Indigenous ways of "seeing, being, and expressing" the environment in today's globalized world. Joni Adamson's recent work also productively demonstrates trans-Indigenous *eco*-sensibilities in Indigenous literature and cinema. For example, in "Cosmovisions: Environmental Justice, Transnational American Studies, and Indigenous Literature" and "Source of Life: *Avatar*, Amazonia, and an Ecology of Selves" (both forthcoming 2013), Adamson articulates the notion of Indigenous "cosmovisions." While Adamson borrows the term cosmovisions from the Universal Declaration on the Rights of Mother Earth, which was framed at the World People's Conference held in Bolivia in 2010, she links the Indigenous concept to multispecies ethnography research and material ecocriticism that also acknowledge the "'pluriform and multivocal worlds' of 'human and natures'" (Walls, 10 qtd. in Adamson, "Source" 7). Adamson's argument, which she has made before in her 2006 *American Indian Literature, Environmental Justice, and Ecocriticism,* is simple yet worth reiterating in "Source of Life's" current grounding of material ecocriticism. Highlighting Juan Carlos Galeano's *The Trees Have Mothers: Amazonian Cosmologies, Folktales, and Mystery* (2008) and James Cameron's *Avatar* (2009) as her case studies, Adamson demonstrates how both films—one an independent documentary and the other a Hollywood blockbuster—encourage Indigenous ways of "seeing, being, and expressing." Specifically, each draws attention to how traditional Indigenous cosmologies can frame contemporary environmental injustices. In centering traditional worldviews in a contemporary context, Adamson argues that the films also recognize modern local and global influences and effects.

Below, I further Adamson and my own arguments. Recognizing the value of Indigenous cosmologies as "theory" that can frame contemporary actions, I focus on Shelley Niro's 2009 *Kissed by Lightning* (*KBL*). I also draw from additional trans-Indigenous scholarship, and personal communication with filmmaker Shelley Niro to articulate the concept of "alternative *eco*-contact" and to substantiate my claim that Fourth Cinema from North America should not be neglected in conversations of the Global South. While the immediate experiences of those in the Global North and South are not necessarily the same, the conversations can prove productive in thinking about trans-Indigenous experiences.

B. "THE BUSINESS OF THE TRICKSTER . . . TO FIND A WAY TO LOOK AT THINGS DIFFERENTLY"

A black screen's momentary silence. Into which blows the low swoosh of wind, itself superseded by susurrations. The screen swells with the rustle of something both natural and manmade: seeds or shells or sand drifting and colliding in a hollowed tube, a rain stick; almost the sound of water. Itself punctuated in equal and distant measure by a single muted drum beat as the film's black screen slowly fills with the following paired lines, each fading in successively:

> The combined confederacy of the Six Nations
> is known as Haudenosaunee.
>
> Haudenosaunee means "People of the Longhouse,"
> or more accurately, "They are building a long house."
>
> The term is said to have been introduced by The Great Peacemaker
> at the time of the formation of the Iroquois Confederacy.
> It implies that the Nations should live together
> as families in the same longhouse.

The first three pairs of lines fade out, while the last remains, finally to dissolve into a medium-long shot of *KBL's* first scene. We find ourselves looking across a mud driveway at a trailer home propped up on stilts alongside a small wooden one-story house. Behind, the woods are bare; in the foreground, at the edge of the driveway, tatters of muddy snow patch the ground. There's a bedraggled feel to the whole scene: dirty snow, muddy driveway, the grubby off-white of the trailer, a plywood board propped carelessly by the wood house, the gray puffs of smoke drifting from the trailer's metal chimney and an abandoned fire pit in front of it.

As the camera lingers, and the mournful pull of a solo viola fills the soundscape, viewers are confronted both by the heaviness of the drab scene and by questions of why. Certainly for this viewer, the trailer home, though recalling the longhouse of the film's opening lines, appears symbolically unmoored, propped up as it is on stilts and centered in the tight framing of the shot. It could literally be anywhere, its flimsy-looking sides and its setting's smallness and untidiness suggesting hardship and poverty replicated across many parts of our contemporary, globalized world, especially in colder climes. In this somewhat ubiquitous image of abjectness, there is, however, one sign of color that anchors us into specificity. Two large, purple flags frame the entrance of the wooden house. They are easily recognized by those familiar with the Iroquois Confederacy comprised of Six North American Indigenous Nations—The Mohawk, Onieda, Onandaga, Cayuga, Seneca and the Tuscarora.

The tension of being unmoored while simultaneously located in an Iroquois frame is reiterated in the shot that immediately follows; a dissolve that pans down the side of a skyscraper to a city street corner, where a Mohawk warrior stands alone, dressed in traditional headdress and garb, and turns to watch a car go by. This scene dissolves into swirling gray clouds, a crack of thunder, and the opening credits of the film that continue as the clouds dissolve into a tracking shot of an interior lived-in space. As a soft male voice echoes into this space, "How did we get lost out here so fast," the camera tracks through the small, dark room full of the knick-knacks of everyday (a cigarette left smoking in an ashtray, a tea pot) and symbols of Indigenous identity (decorative corn, a feathered headdress). The tracking shot is now accompanied by music of the rain stick and single woman's voice singing in her Native tongue.

While *KBL* is billed as one woman, Mavis Dogblood's journey through the grief of having lost her husband, Jesse Lightning, this opening sequence makes apparent a larger commentary on the politics of Indigeneity. As Lawrence Abbot writes about Niro's long portfolio of artistic endeavors, "the interplay of the personal and the political, the mixing of the minutiae of daily life with the broader forces that impact that life is one of the underlying structures of Niro's work" (349). These first two minutes reveal the interplay as Niro adroitly places us at the intersections of the local and global, past and present, rural and urban, nature and culture(s).

The interplay is further strengthened in the last minute of the opening credit sequence, where we are first introduced to Mavis. Framed in medium-shot, she's beautiful; appears in the prime of her life, well-dressed in a contemporary buckskin jacket with traditional beading of colorful flower designs. But she's distraught, screaming into the wind and snow that blows away her words, which are also overwhelmed by the soulful voila. We appear to be in some sort of dream sequence, as the camera cuts to a medium-shot of Jesse, also well-dressed, and playing his viola with the wind and snow swirling around him. As if he's heard Mavis, he stops playing his viola to turn to look her way. As he turns, his image is underlain by another, that of a painted mouth, teeth wide open, screaming. Quick cuts juxtapose these painted underlays below Jesse, then Mavis, then her image dissolves into the underlying painting of a person kneeling in grief. These paintings, we learn soon, are her renditions of the story of Hiawatha and the Peacemaker. While, Jesse had, before his untimely death, been working on a series of musical compositions inspired by the story, Mavis, an artist herself, is working on a series of paintings that tell this important story from her Iroquois past to pay homage to her dead husband and also pay her bills.

At one level the narrative arc of the story is simple: Mavis, driven by the basic logistics of economic survival must make a journey out of her trailer home on the Mohawk reserve to New York City where she is scheduled to exhibit her paintings at an art gallery. On another level, Niro gives us much to think about as Mavis makes her journey across the Canada-U.S. border and through the traditional homelands of her Mohawk people. Most specifically, as Mavis makes her journey, accompanied by her supportive but bumbling friend Bug, we witness her beginning to engage with the world around her such that the film leaves us in a very different mood from the somber, sorrowful one with which it started. This journey is filled with moments of wry comedy, where one can draw from Allan Ryan's well-applied use of a Sharon Manybeads Bowers quote to Niro's work: "The business of the trickster is to take her work very seriously. It is to find a way to change the rhythm, to find a way to look at things differently" (Bowers, 141 qtd. in Ryan 52).

Below, I draw from these moments in *KBL* to suggest how the film helps its audience see Indigeniety differently; that is, not locally marginalized and victimized but globally centered and resilient. In doing so, its use of Iroquois cosmology articulates a natureculture continuum with relevance to contemporary Indigenous socio-ecological realities.

C. "ALTERNATIVE ECO-CONTACT": WHAT THE PEACEMAKER CAN TEACH US TODAY ABOUT TRANS-INDIGENEITY

While its specifics vary from source to source, and its oral telling, like that of most cosmological epics, can take up to ten days,[6] the story of Hiawatha and the Peacemaker explains the origins of the Iroquois Confederacy. It describes the efforts of two men, Hiawatha, a statesmen from one of the Nations (some say Mohawk, some Seneca) and an outsider, Deganawida or the Peacemaker (some say a Huron), who collaborated to broker the "Great Law of Peace" that helped unite the original five nations of the Confederacy, and subsequently the sixth.[7] There is historical evidence to suggest that Hiawatha lived in the sixteenth century (Venables 20; Snow 57). In my conversation with her, artist Shelley Niro explains that the story occurred after "first contact" and the ensuing devastations to Native populations from disease and resource pressures as Europeans began to arrive in the Americas. As Hiawatha and the Peacemaker countered conflict and social upheaval, their legacy lives both in the organizational functioning of the Confederacy they helped create and in legends passed on through Iroquois generations. As many agree, the democratic transnational alliance the two forged was remarkable and proved resilient against the growing onslaught of European influence.[8]

Central to many, like Niro, this message of resilience has "importance to contemporary times" (Niro). Niro states, "There's conflict and grief. How do you get out of the conflict? I want you to survive conflict." Niro sees the Peacemaker's story as an exemplar for overcoming hardship, and hardship is something she is all too familiar with. Elaborating, she says, "I don't know if you know anything about the statistics of Native people, especially Native people living on reserves. The conditions are really dire. You know, a lot of suicides of youth and a lot of tragic deaths, always, so there is this form of hopelessness" (personal communication, April 9, 2013). *KBL* is Niro's attempt to confront this hopelessness "square on" but also in refreshingly creative ways.

In an interview with Cora Bender, about one of her earlier art installation, "The Essential Sensuality of Ceremony," also inspired by the Peacemaker story, Niro states: "Power, peace, and righteousness—those are the three words that every Iroquois person knows—but at the same time when you keep saying things over and over again you become very lazy in your own mind about what it's supposed to mean, or how it's supposed to direct your life" (48). As Ryan's earlier use of Manybeads Bowers' quote and the title of the installation suggests, Niro coaxes her viewers to re-think the meaning of the story by focusing on its dimensions of sight, taste, touch, smell, and sound. She explains, the Peacemaker helped his people eat the right foods, versus being cannibals, helped them use tobacco for peace, and he taught them ceremonial singing. These activities, engage the physical senses necessary for "bringing joy to your life on a daily basis" (49).

Similarly, in *KBL*, Niro helps audiences re-think the meaning of the Peacemaker story through Mavis' growing ability to participate in the material world around her. In traveling from her reserve in Canada to New York City, Mavis is pulled out of her grief-isolated "cocoon" and thus, begins her "healing process." Niro punctuates Mavis' journey with events that take place at tables where food is served. From Mavis' initial inability to partake in the meal Bug prepares and sits down to eat with her at her small trailer table, to the invitation she gets to eat with extended family right before she begins her journey south, to a diner meal interrupted by African-American gospel singers along the route to New York City, to the generous hospitality from Josephine Lightning, Jesse's grandmother who lives along the route, and to the final scenes that capture Mavis a year later, visiting Josephine with the rest of her extended family and Bug, whose child she now bears, Niro uses food as a material prop to trace Mavis's journey out of her inner turmoil and into the social sphere of community engagement and reconciliation. Like for her people from the Peacemaker's time, Mavis' re-learning the ability to enjoy both food and social company demonstrates the inextricable weave of material and mental, nature and culture.

Specifically, the process of experiencing pleasure through one's senses hints at an eco-sensibility (after all, there's a biological wiring that grounds us at the intersection of our material bodies and that of the environmental around). But, this eco-sensibility is not simply personal; Niro expands it to the political, and beyond the local to an Indigenous concept of the transnational by contextualizing Mavis' story in the Peacemaker's story. In particular, *KBL*'s use of the Peacemaker story reframes and challenges traditional Euro-American notions of nation-state, which have repeatedly used forceful coercion in maintaining power.

By turning to a model of Indigenous nationhood that is founded on responsibility and interrelatedness, Niro prioritizes a story that Paul Lai and Lindsay Claire Smith would term "alternative contact–contact apart from narratives of 'first contact' between Native Americans and Europeans" and instead "among Indigenous Americans and other populations in the United States and around the world" (408). Niro referenced "first contact" in her interview with me, but in its many tellings, including in *KBL*, the Peacemaker story, does not evoke this context. Instead, its focus is on Indigenous-to-Indigenous relations through the forming of the Iroquois Confederacy out of five (then six) nations and the presence of an outsider who assisted. In addition, these Indigenous relations are "based on care and responsibility for land that all can share" (Smith 7).

In her interview, Niro emphasizes this sense of a land-ethic as closely woven into the Peacemaker story, explaining that as communities faced with conflict and disease "lost their elders," they didn't have their traditions—"they didn't know how to go through life and not knowing etiquette . . . they didn't know the proper things to eat; the Peacemaker teaches people what they should be eating." In helping the Haudenosaunee re-learn their traditional ways of eating based on *dependence* on the land, not on cannibalism, the story highlights the importance of maintaining the health of the environment for the health of its people.

Two things become evident here. First, interlayering the political contexts of the Peacemaker legend in Mavis' everyday experiences situates the individual in a larger socio-ecological system and second, this socio-ecological system is not indicative of an easy harmony with nature but instead, with its focus on a historical moment of conflict, represents what Jace Weaver has suggested speaks to the "extremely difficult struggles of indigenous people to meet ecological challenges confronting them" (xvi).

While the Hiawatha story is the central guiding motif that speaks to this entanglement of nature and culture, there are other elements in KBL that clue these eco-sensibilities of natureculture. Most obviously, the film's title evokes nature in its reference to lightning and its repeated visual motifs of gray clouds, rain, and falling snow (as seen in the opening sequence). Niro explains her allusion: "I wanted the word lightning to be seen as powerful but beautiful; kind of dangerous. But then, Jesse Lightning is also literally killed by lightning."[9] We see a tension in lightning's beauty and danger; we also see the tight amalgam of lightning as a human (Jesse Lightning) *and* as natural elemental force. This entanglement is reflective of traditional Iroquois understanding of socio-ecological complexity. As Dean R. Snow, referencing Iroquois origin stories of Sky Woman's twins, Sapling and Flint, writes:

> The co-existence of good and evil pervades the Iroquois cosmos . . . The wolf is not pervasively evil, though he may appear to be to the rabbit; and the rabbit is not all good, for he destroys the young shoots of Iroquois crops" (4–5).

While Jesse Lightning is struck by lightning and his death has struck Mavis to grieve inconsolably, Niro alludes to how nature is neither good nor bad, but an essential presence in *KBL*'s world: "Snow comes and it covers everything up and makes everything beautiful" but it's also "cold." Snow mirrors Mavis' internal state of grief as she cocoons herself alone in her trailer home and is unable to partake of the pleasures of daily senses. In her dreams of Jesse, like the one from the opening sequence, she screams but without sound; she reaches for Jesse, but is unable to touch him. In her day to day activities she has also closed herself off to Bug's gentle overtures as well as interactions with extended family—like with Jesse's son, Zeus (from his first marriage), and Jesse's perky ex-wife, Kateri, who live next door. Yet, the visual beauty of Mavis' dreams seem to signal that grief is also a vital aspect of life.

Niro states, KBL is not "an environmentally conscious film" in any obvious way but it hopefully presents nature as more than metaphor. She explains: "it's almost as if there are two parallel realities, side by side, or one on top of the other. One involves living today in a sort of industrial world. And the other involves living in a spiritual world" (personal communication, April 9, 2013). In both worlds, material and spiritual, nature is enmeshed into the cultural landscape in ways that acknowledge its vital presence yet don't explicitly offset it.

This insight of natureculture's continuum—that nature is vitally enmeshed, though *not* necessarily prioritized—is an important characteristic of *KBL* as a narrative of alternative eco-contact. And, it is well worth considering the potentials such a narrative has to connect contemporary Indigenous eco-experiences across the globe. On the one hand, such a narrative forwards a worldview that is a common feature of many Indigenous spiritual traditions practiced across the globe, and thus may be easily relatable. More importantly though, as more and more contemporary Indigenous peoples across the globe find themselves distanced from traditional natural contexts (for example, of subsistence farming, or hunting and gathering), such narratives of alternative eco-contact can speak empathetically to their everyday realities where nature has been subsumed into the material goods, financial survival, and technological artifacts that embody their cultural landscapes. In such circumstances, these narratives can assist in helping one choose how to engage with this world's enmeshed contours, as is the case with Mavis. Such thought, with its acknowledgment of socio-ecological complexity and nuance can be crucial to our ability to address socio-ecological crises and to "live together, as families in the same [Earthly] longhouse" (Niro, *Kissed by Lightning* epigraph).

KBL's use of the Peacemaker story not only encourages Mavis to think about how she might live, mentally *and* materially, again with her family, but it also provokes us to consider how we might live with others who occupy our global contexts. Niro uses Mavis' Canada-U.S. border crossing experience to creatively comment on this larger political possibility. In doing so, she addresses another vital dimension of narratives of alternative contact, the ability to connect "with Indigenous peoples of other continents and the struggles of migrant and minority communities more broadly" (Lai and Smith 408). As I discuss below, *KBL* also makes the ecological dimension of these connections apparent.

About thirty-five minutes into the film, and early in the journey to New York City, we find ourselves with Mavis and Bug at a Canada-U.S. border check point. They, like others, wait in their car, watching through the swish of wipers brushing falling snow off their windshield, as their line slowly inches forward. Bug, who is hurt and bandaged up from an earlier fall on the ice is encouraged by Mavis to take his painkillers as they wait in a "line that goes on forever" (Mavis's words). As they inch forward, Mavis says to Bug, "Whatever you say, don't say you are Mexican."

These words create an interesting resonance. They seem both out of place (after all, so far there's been nothing to signal attention to Mexicans or Mexico) and very relevant to the particular situation in which Mavis and Bug find themselves. Almost anyone from the Global South is well aware of the added scrutiny that such allegiance engenders at border crossings, especially into the Global North. Niro plays with this tension to signal a connection between the North American Indigenous experience and one encountered all too frequently by those from the Global South. As Mavis pulls up to the check point, the irritable face of a Causacian customs officer looms in her window, demanding citizenship. She responds "Six Nations." Bug, drugged and drowsy from his painkillers, accidentally says "Mexican" though he too, as we have seen so far, is Six Nations. The customs officer demands they show their passports and come inside for further questioning.

Inside, Mavis finds herself seated next to another Iroquois person. He asks conversationally, "Do you always get stopped?" When Mavis states that she does not cross the border frequently enough and that this is her first time being stopped, he responds, "They are always looking for something" despite the various treaties that are supposed to allow free movement of the Iroquois across their ancestral territories. This "always looking for something" is reiterated through Mavis's subsequent exchange with the customs officer, who fires a list of questions at her such as "what's in the van?" And, whether her paintings are "pornographic?" Or, if she's carrying "teepees and tomahawks?" And, if she's got "documentation" to verify her claim that she's going to be showing her paintings in New York. As Bug sprawls asleep on a bench, the customs officer also asks, "Do you have prescriptions for your companion's medications?" Niro furthers the hierarchical play of power through her placing of the characters in mise-en-scene. Mavis sits on a hard bench in the impersonal space of a waiting room while the officer towers over her, impatient in his gestures and facial expressions.

As the man waiting with Mavis seems to have indicated, the customs officer's belittling attitude is an all too familiar occurrence for Iroquois seeking to cross the border. But, we can also credit Niro for re-framing this confrontation beyond the local experience to a global one through her inclusion of the Mexican reference. While such transnational engagement is evocative, so is Niro's gentle humor as the scene plays out through Bug's drugged-out cluelessness, the pleasant geniality of the man also waiting with Mavis, and Mavis' own bemused surprise at the customs officer's questions. It's almost as if Niro isn't interested in generating outrage at a Western society that has imposed its arbitrary rules on Indigenous and Global South citizens as much as she wants to remind us of both the inhumanity *and* humanity in such everyday occurrences. This humanity is hinted at as the scene ends with the customs officer leading Mavis and Bug back to their car. He even opens the passenger door for Bug, and as the camera pulls back into a long shot of him standing in the snow waiting for Mavis to pull away, we are presented not with an image of racist or "invader" hubris but one of an ordinary individual caught in a moment of everyday vulnerability. He's a man standing in the cold snow.[10]

Through this subtle movement, Niro re-orients us once again into the mesh of Iroquois cosmology's natureculture. A point driven home further by the ensuing landscape shots that follow Mavis and Bug's drive into New York State, the ancestral homelands of the Mohawk people. Accompanying this portion of the drive is an upbeat tune fusing Mohawk lyrics, percussion, and Western instrumental beats. The sun glints off the snowscape, water sparkles in its light, and Bug, now revived, reminisces about taking a bus trip with his mother and sister "in the Peacemaker's footsteps." This is when he states, "When your feet touch the Earth, your soul mixes with the souls of your ancestors." Importantly, such a belief shifts *KBL* beyond a story simply of alternative contact, which re-centers attention on Indigenous peoples as active agents not merely victims of colonization, into that of alternative *eco*-contact. Earth and its tactile presence, whether in the snowy landscape, or the more globally felt phenomena of a crack of lightning, the taste of food, the touch of sun, or the earth beneath one's feet, are essential to our existence—individual or collective. It is the interwoven fabric of reasons we have "to live and make better humans of ourselves."[11]

D. FOURTH CINEMA: EMBODYING OUR EARTHLY LONGHOUSE

The more one engages with Shelley Niro's *KBL*, the more one is made aware of her artistic prowess in using Iroquois cosmology to ground an Indigenous present. As the sampling of scenes above suggest, nature is a vital component of *KBL*'s cultural landscape. It is part and parcel of the transnational and transcorporeal experience of Niro's characters who not only have to cross literal borders to traverse their ancestral homes but who also engage with the mental *and* material ramifications of historical legacies, both debilitating (as in acts of dispossession and loss—collective and individual) and inspiring (as in the Peacemaker story). Niro's emphasis on hope out of "hopelessness" is solidly anchored in this mesh of natureculture. Not only does her protagonist, like the Iroquois people did, transition out of sorrow and re-learn the ethics of individual *and* communal care by beginning to partake in the physical senses that "bring[ing] joy to your life on a daily basis" but in doing so, Niro, through the cinematic apparatus, extends the possibility of this experience to her audience.

As the average moviegoer knows through experience, and as ample research in cognitive film theory acknowledges, cinema itself triggers a transcorporeal yet *embodied* response (Noel; Plantinga and Smith; Shaw; Grodal). In effect, as audience members, we experience pain and pleasure vicariously, often viscerally cringing at physical threats or laughing out loud with those on screen. Through its aural and visual cues, *KBL* cues its viewers emotionally and cognitively towards resilience and reconciliation—the soulful viola gives way to an upbeat contemporary Mohawk tune, or the threat of Caucasian-settler power dissipates into gentle comedy. While our own material and cultural lenses can, and do, influence how we receive these cues, and whether we see our own responses as enmeshed natureculture, bringing an Iroquois way of "seeing, being, and expressing" to *KBL* can enable this latter understanding.

Such an Indigenous-centered approach can speak across and beyond the Iroquois experience in potentially meaningful ways. Specifically, when we consider *KBL*'s presence at Indigenous film festivals, including at the Smithsonian's "Mother Earth in Crisis" film festival, where it screened alongside films produced in countries more easily recognized as the Global South, we might recognize it as a complement to those films that speak more explicitly to ecological crises. Its message of hope and resilience can be a powerful force in the face of socio-ecological adversity. And, its narrative of alternative eco-contact is a necessary first step in comprehending Indigenous eco-experiences beyond those of land claims and the settler-Indigenous binary. Such narratives are also valuable in reminding us of how trans-Indigeneity can flow outwards to the non-Indigenous as they encourage us to see how we are all kissed by nature in its enmeshed continuum with culture. To recognize this intimate entanglement is imperative if we hope to live sustainably together as families in the same Earthly longhouse.

ACKNOWLEDGMENTS

I am especially grateful to artist Shelley Niro for speaking with me about her film and her work. I am also grateful to Elizabeth Weatherford at the Smithsonian Native Film + Media Center for her assistance with the logistics of this project.

WORKS CITED

Abbot, Lawrence. "Interviews with Loretta Todd, Shelley Niro, and Patricia Deadman." *The Canadian Journal of Native Studies* 16.2 (1998): 335–73. Print.

Adamson, Joni. "Cosmovisions: Environmental Justice, Transnational American Studies, and Indigenous Literature." *The Oxford Handbook of Ecocriticism*. Ed. Greg Garrard. New York: Oxford UP, 2013. Print.

———. *American Indian Literature, Environmental Justice, and Ecocriticism: The Middle Place*. Tucson: U of Arizona P, 2001. Print.

———. "Source of Life: *Avatar, Amazonia*, and an Ecology of Selves." *Material Ecocriticism*. Ed. Serenella Iovino and Serpil Oppermann. Bloomington: Indiana UP, 2014. Print.

Alaimo, Stacy, and Susan Hekman, eds. *Material Feminisms*. Bloomington: Indiana UP, 2008. Print.

Alaimo, Stacey. *Bodily Natures: Science, Environment, and the Material Self*. Bloomington: UP, 2010. Print.

Barclay, Barry. "Celebrating Fourth Cinema." *Illusions* 53 (Winter 2003), reprinted Word Document. Hawaii Public University. Web. October 22, 2012.

Carroll, Noel. *The Philosophy of Horror, Or Paradoxes of the Heart*. London: Routledge, 1990. Print.

Chadwick, Allen. *Trans-Indigenous Methodologies for Global Native Literary Studies*. Minneapolis: U of Minnesota P, 2012. Print.

———. "A Trans*National* Native American Studies? Why Not Studies that are Trans-*Indigenous*?" *The Journal of Transnational American Studies* 4.1 (2012). Web. February 8, 2013.

Columpar, Corinn. *Unsettling Sights: The Fourth World on Film*. Carbondale: Southern Illinois UP, 2010. Print.

Cruikshank, Julie. "Listening for Different Stories: Indigenous Knowledge and its Transformations in a Rapidly Changing North." Association for the Study of Literature and Environment symposium. Juneau, Alaska. June 14, 2012. Keynote Address.

First Peoples: The Legends. "Hiawatha the Unifier." n.d. Web. March 7, 2013.

Ginsburg, Faye, Lila Abu-Lughod, and Brian Larkin, eds. *Media Worlds*. Berkeley: U of California P, 2002. Print.

Grodal, Torben. *Embodied Visions: Evolution, Emotion, Culture, and Film*. New York: Oxford UP, 2009. Print.

Hampton, John. "Unsettling Sex." *ImagineNATIVE* Program, October, 2012. Web. October 18, 2012.

Huang, Hsingya, et al. "Charting Transnational Native American Studies." *The Journal of Transnational American Studies* 4.1 (2012): 1–15. Web. February 8, 2013.

Iroquois Indian Museum. New York, 2013. Web. March 7, 2013.

Iovino, Serenella, and Serpil Oppermann. "Theorizing Material Ecocriticism: A Diptych." *ISLE: Interdisciplinary Studies in Literature and Environment* 19.3 (Summer 2012): 448–75. Print.

Kissed by Lightning. Dir. Niro, Shelley. Perfs. Kateri Walker, Erik Schweig. VTape, 2009. Film.

Lai, Paul, and Lindsey Claire Smith. Introduction. *American Quarterly* 62.3 (September 2010): 407–35. Print.

Lewis, Randolph. *Alanis Obomsawin: The Vision of a Filmmaker*. Lincoln: U of Nebraska P, 2006. Print.

Marrubio, Elise, and Eric Buffalohead, eds. *Native Americans on Film: Conversations, Teaching, and Theory*. Lexington, Kentucky: University Press of Kentucky, 2012. Print.

Monani, Salma, and Miranda Brady. "ImagineNATIVE 2012: Ecocinema and the Indigenous Film Festival." Unpublished manuscript.

Monani, Salma. "Eco-Activism in Indigenous Film Festivals: The 2011 Native Film + Video Festival." *NECSUS: The European Journal of Media Studies*. 1.3 (2013). Web. June 3, 2012.

Murray, Stuart. *Images of Dignity: Barry Barclay and Fourth Cinema*. Wellington, Aotearoa, New Zealand: Huia. 2005. Print.

Oneida Indian Tribes of Wisconsin. "The Great Law of Peace." 2013. Web. March 7 2013.

Niro, Shelley. *Phone Interview*, April 9, 2013.

The People of Kattawapiskak River. Dir. Obamsawin, Alanis. National Film Board of Canada, 2012. Film.

Plantinga, Carl, and Greg Smith, eds. *Passionate Views: Film, Cognition, and Emotion*. Baltimore: Johns Hopkins UP, 1999. Print.

Raheja, Michelle. "Reading Nanook's Smile: Visual Sovereignty, Indigenous Revisions of Ethnography, and *Atanarjuat* (*The Fast Runner*)." *American Quarterly* 59.4 (2007): 1159–87. Print.

———. *Reservation Reelism: Redfacing, Visual Sovereignity, and Representations of Native Americans in Film*. Lincoln: U of Nebraska P, 2010. Print.

Ryan, Allan. "I Enjoy being a Mohawk Girl: The Cool and Comic Character of Shelley Niro's Photography." *American Indian Art Magazine* (Winter 1994): 44–53. Print.

Shaw, Daniel. *Film and Philosophy: Taking Movies Seriously*. London and New York: Wallflower Press, 2008. Print.

Smith, Andrea. "American Studies without America: Native Feminisms and the Nation-State." *American Quarterly* 60.2 (2008): 309–15. Print.

Snow, Dean R. *The Iroquois*. Oxford, UK, and Cambridge, MA: Blackwell, 1996. Print.

Venables, Robert W. "Introduction." *The Six Nations of New York: 1892 United States Extra Census Bulletin*. Ithaca: Cornell UP, 1995. vii–xxv. Print.

Wood, Houston. *Native Features: Indigenous Films from Around the World*. New York and London: Continuum, 1995. Print.

NOTES

1. Alanis Obomsawin is perhaps Canada's most prominent First Nations documentarian. She has made over thirty films with support from the National Film Board of Canada and is well known for her quiet and unflinching representation of First Nations experiences. Randolph Lewis' treatise, *Alanis Obomsawin: The Vision of a Native* Filmmaker (2006) is an excellent place to begin to engage with her work.

2. While Stacy Alaimo's coedited (with Susan Hekman) *Material Feminisms* (2008) and her *Bodily Natures: Science, Environment and the Material Self* (2010) are well referenced in the ecocritical literature, Joni Adamson's "Source of Life: *Avatar, Amazonia*, and an Ecology of Selves" (forthcoming in the edited volume *Material Ecocriticism*, 2013) also draws on anthropologist Julie Cruikshank's keynote symposium address

"Listening for Different Stories: Indigenous Knowledge and its Transformations in a Rapidly Changing North." In all these works, including the recent *ISLE* journal cluster (2012) edited by Dana Philips and Heather Sullivan, Donna Haraway's concept of natureculture is productively applied. As Serenella Iovino and Serpil Oppermann write in their ISLE essay, "we can talk of matter and bodies as combinations of material-discursive or 'material-semiotic agencies'" (450).

3. In their edited issue "Alternative Contact: Indigeneity, Globalism, and American Studies," Lai and Smith remind us of the limited ways in which North American Indigeneity has been traditionally perceived within the binary logic of settler-to-Indigenous. These representations have not only stereotyped Indigenous people as "static" and stereotyped and but also often limited Indigenous eco-narratives to those of concerns for land claims (408).

4. Italics are from the original.

5. Raheja's *Reservation Reelism* works well to show how Natives have subverted such hegemonies when working with Hollywood. See also Faye Ginsburg, Lila Abu-Lughod, and Brian Larkin's edited *Media Worlds* for additional discussion of these complexity (2002).

6. See for example, the websites of the Oneida Indian Tribes of Wisconsin; First Peoples: The Legends; and The Iroquois Indian Museum.

7. The Tuscaroras, an Iroquois language group that had settled in the Carolinas, had suffered from the influx of colonizers and consequently many fled back north to Pennsylvania and New York. They were assisted by the Oneida and officially became the sixth nation of the Confederacy in 1722.

8. There are suggestions that this organization was thought-provoking to founders of the United States of America, such as Benjamin Franklin, who met with Iroquois chiefs in the 1700s. See Robert Venables' Introduction to *The Six Nations of New York*, which includes an excerpt quoted from a personal letter by Franklin (xi).

9. Niro explains that the film's budget didn't allow for Jesse's death by lightning to be explicit on screen but she confirms that he is indeed *literally* "kissed by lightning."

10. Niro explains that the "Mexican" reference and the exchange with the customs officer are inspired by real-world incidents encountered by herself and family and friends. She also states that in this scene, just as in other ones (for example, a wonderful one of meeting African-American gospel singers in a small-town diner full of Caucasians that look suspicious of these people of color), her point is not to delve into settler-Indigenous animosities. Explaining the border officer presence, she states, "some of them are just so hard. And you are ready for them; and then they turns out to be the nicest guy in the world."

11. I place the phrase in quotes as it draws directly from Bug's words, which describe the Peacemaker giving his people "reasons to live and making better humans of themselves."

TEN

"Under all the laws, natural, human, and divine"

Reinterpreting La Leyenda Negra's Colonial Purpose[1]

Dora Ramirez-Dhoore

With reference to one of the most elusive legends affecting how Latin American space is perceived in the world and the anti-Hispanicism found in our current society, La Leyenda Negra (the Black Legend) has affected the manner in which the Spanish, and consequently Latinos, have been perceived in Europe and the Americas since the conquest of the New World. The Black Legend's beginnings are attributed to the bishop Bartolemé de Las Casas (1522), who wrote extensively and negatively about the Spanish Conquest of the New World and had great influence in pushing the Black Legend's agenda (not without due cause in giving voice to the indigenous populations of Latin America). Yet, the Black Legend can be roughly defined as a "reaction to the large body of opinion which actually holds that the Spanish are inferior to other Europeans in those qualities commonly regarded as civilized" (Maltby 3). In an ironic turn, the concept of inferiority has often been placed on those colonized by the Spanish—the indigenous people—and influences the dichotomous rhetoric that allows for the colonization of the Global South to continue. The discourse used in the Black Legend is worthy of study as it redefines the Global South for purposes of colonization and exploitation—of the people, the land, and the economy. Walter D. Mignolo points to the connection between race, imperialism, and the Black Legend in the afterword to *Rereading the Black Legend: The Discourses of Religious and Racial Difference in the Renaissance Empires* (2007):

> Racism emerged as a discourse to assert the superiority of Western Christians and as justification for land appropriation and exploitation of supposedly lesser human beings. Today racist discourse has a similar function in keeping Chinese and Iranian expansion at bay and criminalizing immigration in the United States and in Europe. The Black Legend is a piece in the puzzle of Renaissance discourse on race that put forward the imperial difference among European powers. (314)

The discourse that makes up the Black Legend allows for difference to be the reason for a continual colonization of the Global South.

La Malinche is an offshoot of the Black Legend and is considered to be a national symbol, reflected in Mexico's national cry, *Viva Mexico! Hijos de la Chingada!*[2]—calling up the moment of Spanish conquest. Her figure is negatively framed as la Chingada (the whore)[3]—the Other. She represents conquest, sexuality, power, national identity, and indigenous memory. I argue that the Black legend and the legend of La Malinche work together in maintaining the perception that Mexicans and their descendants are a raped nation of people, and I connect this to the silenced voices of indigenous populations and the continued destructive use of the physical land in the Global South. This sentiment extends into contemporary notions of biopower and social control of the Mexican people with Hernán Cortés as the image of power and privilege and La Malinche as the manipulated body and country. These legends work to sponsor continuous colonization of a people and their environment; thus this essay reworks the figure of La Malinche while reinterpreting the rhetoric of colonization and racism in the Black legend to illustrate how these legends work in tandem. Laura Esquivel's novel, *Malinche* (2006), helps in illustrating the effects colonialism has had on the geographical and religious landscapes of the conquered populations of Mexico.

Central to this argument is how the discourse that has fueled land exploitation, genocide, and political issues can be used as a methodology in reading literature. To do this, I focus on the water and land imagery found in Esquivel's *Malinche,* which illustrates how the Black Legend continues to redefine Mexico's environment as both usable and expendable. This essay engages a history that separates indigenous Mexico from the rhetoric of the Black Legend in order to change perspectives about colonization, indigeneity, and land use, and to enter a fuller understanding of what conquest means in history when politics are connected to the Earth.

REVISITING LA LEYENDA NEGRA

The Black Legend has its known beginnings in 1604 when the satirist Francisco Gómez de Quevedo y Villegas first mentioned it in "España Defendida," and when it was revisited by Julián Juderías in 1912 in his book *La leyenda negra y la verdad histórica* (*The Black Legend and Historical Truth*). As a Spaniard himself, Juderías points out how the treatment of the Spanish in the foreign press negatively characterized the Spanish: "In short, we understand as the Black Legend the legend of inquisitorial Spain—ignorant, fanatical, incapable, now as in the past, of being considered among the civilized nations, since we always prefer violent repression and we are the enemy of progress and of innovations" (Powell 11). This statement makes apparent that the Black Legend works off the civilized/savage binary.

The essentialist attitudes attributed to the Black Legend affect the way literature is produced and consumed. Rafaela G. Castro summarizes La Leyenda Negra most succinctly in Chicano Folklore:

> The Black Legend originated in the sixteenth century during the Protestant Reformation, with the beginnings of a rejection of Catholicism and abhorrence of the political and military power of Spain. When Spain and Portugal remained loyal to Catholicism, and expanded their empires by settling in South America, Central America, the Caribbean Islands, and parts of North America, the rest of Europe looked on very disapprovingly. This viewpoint has continued to color everything about it since then, including the grim legacy left by Spain in the New World. It has justified the dictatorial treatment of Latin America and Mexico by the United States since the early nineteenth century. Biased attitudes toward Catholic Spain were inherited by Anglo Americans and brought to the New World colonies by Protestant English settlers. (137)

In the introduction to *The Devastation of the Indies: A Brief Account* (1974), Bill M. Donovan poses a few questions to the reader. He asks, "What is the proper moral reaction to monstrous injustice?" and continues by stating that "*The Devastation of the Indies*" raises the profound question whether something intrinsically immoral in the West's ethos has underlain all Western/non-Western relations from the earliest voyages of discovery" (qtd. in Casas 23). In his discussion, Donovan continues: "The Black Legend survives in part because groups such as the Puritans never truly questioned the basic character of European/Indian relations" (qtd. in Casas 23). In this way, nationhood could be determined to be rigid patriotism. Michel de Montaigne (1580) would agree, stating:

> I find that there is nothing barbarous and savage in this nation, by anything I can gather, excepting, that every one gives title of barbarism to everything that is not in use in his own country. As, indeed, *we* have no other level of truth and reason, than the example and idea of the opinions and customs of the place wherein we live: there is always the perfect religion, there the perfect government, there the most exact and accomplished usage of all things. (66 italics mine)

Four centuries later, in his 1978 book, Edward Said attributes this attitude to Orientalism and states, "Indeed, my real argument is that Orientalism is—and does not simply represent—a considerable dimension of modern political-intellectual culture, and as such has less to do with the Orient than it does with '*our*' world" (12, italics mine). In *The Tree of Hate* (1971), Phillip Wayne Powell calls for the "Destruction of the Black Legend and its long chain of echoes and consequences—that historical 'tree of hate' whose fruit poisons the English-speaking world and robs it of the capacity to approach the Hispanic world with fairness, with sympathy, and without prejudices—must be the first great step in bridging the chasm that now separates the two largest cultures of the West" (167). In *The Black Legend in England* (1971), William S. Maltby explores the legend's far-reaching effects on current views of Latinos in the United States: "The 'Black Legend' has undeniably played a large part in the troubled relations between the United States and its Spanish-speaking neighbors, and it continues to influence British policies with regard to Spain. However difficult it may be to isolate, it remains a factor in international affairs" (9). In "'Race' and 'Class' in the Spanish Colonies of America" (2007), Yolanda Fabiola Orquera explains how the Black Legend is found within scientific reasoning illustrating how colonization continues today: "In the case of the conquest of the Americas, Spain imposed its cognitive paradigm over local preexistent communities, using 'scientific' reasons to refuse the validity of their forms of thinking and converting them into residual knowledge" (167). These critical voices reveal the dominance of rhetoric that segregates the Spanish-speaking world from the English-speaking world. This dominance also reflects the attitudes imposed upon the people, the land, and the environment more generally. These scholars point to the necessity of questioning relationships and recognizing interconnectedness. The legend of La Malinche, when viewed through the Black legend, symbolizes this divided rhetoric that maintains rigid nationalistic attitudes and segregates relationships.

REDEFINING THE NEW WORLD

In connecting Montaigne's *we* and Said's *our* with nationhood, it is important not only to revisit the figure of La Malinche[4] but to disengage her from the rhetoric of conquest that has defined her. Octavio Paz defines her as a negative symbol for the country. He states: "Her passivity is abject: she does not resist violence, but is an inert heap of bones, blood, and dust. Her taint is constitutional and resides . . . in her sex. This passivity, open to the outside world, causes her to lose her identity: she is the *Chingada*" (208). Paz's words frame la Chingada as the Other and does not move La Malinche outside of a negative portrayal. This negative portrayal of Malinche has been repudiated by feminist scholars who re-imagine her through the use of new symbols of representation which emphasize issues like conquest, sexuality, power, language, national identity, and the subaltern.

The story of La Malinche contains an embedded history of the conquest of land and people. In "Colonial Borders, Native Fences: Building Bridges between Indigenous Communities through the Decolonization of the American Landscape" (2012), Sandra M. Gonzales discusses the exploitation of natives who were gathered into controlled communities specifically as a labor source for mining (314). She connects this labor source to the settlement and conquest of the land by the Spanish intermarrying with the indigenous women. She writes: "However, with few Spanish women, the conquistadores were forced to 'mejorar la raza' or improve the race of Indians by raping native women and taking them as concubines" (314). Thus sex was used as a settlement tool. The connection between the colonization of the indigenous people and the acquiring of land was solidified by redefining the make-up of the ruling class. Gonzalez explicates: "In Mexico, *mestizaje* originally began by taking and raping, or attempting to biologically change the racial composition of the Indian population. Later, *mestizaje* morphed into a dubious, caste-imposed concept based on skin color as opposed to racial mixture" (314). This offered the conquerors an indigenous labor force.

Cortés's position as a prominent Spanish conquistador plays a part in the national mythology surrounding the legend of La Malinche. From these acts of settlement and conquest arrive the son, Martín Cortés, or the first mestizo. Sandra Messinger Cypess writes of other first mestizos who did not receive this title. She points to the conquistador, Gonzalo Guerrero, who was shipwrecked on the Yucatán peninsula and fathered children with his Mayan wife. When Cortés offered to rescue him, Guerrero pointed to his offspring and told him, "'Don't you see how lovely they are.' [. . .] These children could well have become the symbols of the Mexican mestizo, instead of the son of Malinche and Cortés, who arrives chronologically later but remains to this day the icon of mestizaje" (21). What is apparent in this use of legend as national history is the manner in which stories are constructed to fit a message. Cypess acknowledges this:

> I would like to suggest that the preference of Cortés and Malinche's son Martín as *Mestizo #1* reflects the same ideological decisions that erased any other couple from consideration. Issues we can identify today that relate to gender, ethnicity, religion, and class are all relevant. The social standing of Martín's father, as opposed to Guerrero's position, has determined Martín's selection as number one mestizo. He is first in social rank as the son of a conquistador, and his Catholic religion, dress, and journey to Spain, all prove European male dominance over Amerindian women and European cultural dominance over Amerindian ways of life. (21–22)

La Malinche's identity is constructed through her role as a slave, a raped woman, and then by her ambiguous loyalties to her people. After Malinche is sold to Cortés, she becomes his interpreter and mistress, solidifying her historical legacy as mother to the first mestizo. Cypess writes of her as an "exile in her own land" and as "a central figure in the Conquest, an emblem of Mexican national identity and also as a symbol of all Latin American women" (14). The combination of this story working alongside the Black Legend illustrates how social capital allows the stronger, the conqueror, to define the Other.

EXTRACTING INFORMATION

The Black Legend and the legend of La Malinche work together in maintaining notions of biopower and social control of the Mexican people with Cortes as the image of power and Malinche/Mexico as the manipulated body. These attitudes work to continuously colonize a people and their land. Las Casas understood this when he wrote:

> . . . that both natural and human and divine right are something quite different, that they have acquired nothing when the Indians are terrorized into giving submission and tribute to the foreign King but have earned the punishment of the devil and the eternal fires of hell. They do not see that they are even committing offenses against the King of Castile in destroying these kingdoms and in annihilating the inhabitants. Yet such services the Spaniards in the Indies have rendered to their King and are still rendering today. (65–66)

Las Casas sees the precursor of the Black Legend mentality when he writes of the extermination of people and the countless lands various captains laid to waste in the name of the Spanish crown (66). Esquivel also illustrates how the colonization and destruction of these lands has continued to affect the indigenous population of Mexico. Malinalli hears the cries of Cihuacóatl and understanding that the lands will be unprotected, foresees the razed future of her land not long after she is sold to Cortés. Malinalli recognizes that "Without her care what would become of her cornfield? She wouldn't know. From this day forward, she would begin to journey through a path that she had traveled before: being separated from the earth she had grown attached to" (Esquivel 18). Symbolism works within this structure of time and consequence. The Black Legend makes the patriarchal version of the Malinche legend stronger because it gives legitimacy to the use of La Malinche (and la Llorona) as a symbol of Spanish atrocities, through the rape of a nation. Inserting the image of Malinche also helped to portray the Spanish as violators and conquerors, thus adding credence to the rhetoric of the Black Legend. Cordelia Candelaria writes of historical moments that illustrate Malinalli's importance to Cortés, complicating, but not ignoring, the idea of rape of both woman and land as a trope in the legend:

> Of prime importance, moreover, are the recorded examples of Cortés' esteem for La Malinche apart from the understandable, if lamentably chauvinistic, failure to acknowledge a "heathen consort" in letters to the Sovereign. First Cortés named their illegitimate son after his own father, Don Martín, an act of considerable magnitude among the custom-oriented Europeans of that time. Second, he assigned several partitions of land to her during his administration of Mexico. That too was an important deed since she was not a member of the Spanish ruling class which was automatically entitled to property ownership in post-Conquest New Spain. (4)

Thus, the conquered land Malinalli receives from Cortés is symbolic of his conquest and her rape in the legend.

Candelaria furthers her argument by focusing on the image Cortés's contemporaries had of Malinalli, presenting another of the many rhetorical uses this myth has had in slandering the Spanish. She states that Don Fernando de Alva Ixtlilxochitl,[5] "perhaps . . . in writing of his people's role in *la conquista,* was simply reacting against the Spaniard's ultimate blasphemy: allowing a woman into a male-dominated public sphere" (5). In contrast, Bernal Diaz de Castillo "introduced La Malinche's greatness to the world" (5). Candelaria's focus on these men illustrate that the rhetoric used to describe Malinalli's involvement with the conquest was held suspect by a patriarchal system and was thus constructed to fit into that paradigm by either silencing her participation or alluding to the inherent weakness of the Spanish in needing a woman in order to prove successful in their conquest of the New World.

In the novel, *Malinche,* Cortés's character is constantly learning about the indigenous culture from Malinalli, but selectively educating her as well. In his conversation with Malinalli about her religion, he drives her to tears when he tells her not put faith in her gods. As she retorts, she talks of the water and the treasure produced in deep caves. The conversation moves from spirituality to materiality when Cortés questions, "Really? What treasures? Where are these caves?" (64). His reaction proves Cortés's interest in Malinalli as a source of knowledge about the natural resources in the area. Malinalli is bothered and hesitant in answering his question because, "It proved that Cortés was not interested in talking about his religion, or his gods, or his beliefs, or even about her. It was clear that he was only interested in material treasures" (64). She then goes to weep by the river, illustrating her faith in her indigenous gods.

Cortés's attitude was also marked by the Black Legend. In "Nationalism, the Black Legend, and the Revised Spanish Tragedy," Eric Griffin reminds us that "During Elizabeth's turbulent final decade English public culture had been inundated with Hispanophobic stereotypes mobilized again and again in propagandistic efforts to affirm the twin pillars of national sovereignty and the Protestant faith" (357). Cortés's thirst for treasure and gold was used by the English as a method of working against Catholic Spain. Thus, the connection between land, conquest, and rape is used within both legends, illustrating the connection between the Old World and the Global South in matters of colonization. Both legends remove the inhabitants and culture from the geographical and economic, thus imagining the Global South as a primitive territory needing "improvement" through colonization.

LAND, CONQUEST, AND RAPE

The theme of rape also exemplifies the disconnection Cortés had with the environment of New Spain—both spiritual and geographical. Malinalli's religion connects the land, water, and the womb to her deities. She questions Cortés's God as she begins to question her role as his interpreter within the power systems she had experienced within her life as a young female slave. She equates Cortés to the god Ometéotl, who represented duality and the dual nature of the living universe (Esquivel 62). In reference to this, she wonders who the Catholic God's wife is, of which Cortés answers that He has none. The concept of "Without a womb there is no god" (62) creates a division between their worldviews because Malinalli's worldview embraces an inclusion of interconnection, while his is seemingly material. Cortés makes clear their difference in worldviews, when he ends their sexual intercourse "because he was frightened by that sensation of complicity, of belonging" (63). It is in this scene that Cortés realizes that he needs Malinalli in order to learn about New Spain and its inhabitants. Without her he has no access to power. He had much knowledge to learn from Malinalli, which he takes as part of his conquest of the land. She, for him, becomes only part of the landscape of conquest, where her worldview envisions Cortés as her counterpart. Her worldview would assert that without Cortés there would be no Malinalli, while his worldview leaves her as disposable.

Esquivel works to show how the Spanish conquest disconnected Malinalli from her beliefs in divine unity and how Catholicism was built up by negating the deities of the indigenous peoples. Ironically, it is within this scene, where Malinalli tries to understand Cortés's religion, which makes him uncomfortable, because she realizes that his lust was with the material fortune to be found in Mexico. In speaking about human sacrifice, she begins to reinterpret the Catholic God and attempts to tie Him into her spiritual connection to water by believing in Him and by seeing Him in her tears. In reference to Genesis, and in respect to her belief in the necessity of a womb for life to be balanced, she questions Cortés's religion: "There is life that is born without a womb, but it does not remain long on the earth. What is engendered in darkness, however, in the profundity of caves, like precious gems and gold, lasts much longer" (62–63). Malinalli's connection to the God of Water (Tláloc) is part of the narrative structure in her speech and again when Cortés rapes her in the river and rain, while extracting knowledge. Cortés sees Malinalli as "his true conquest" (77); and it is this moment that centers the readers' attention on rape, and what water and the land symbolize. In this scene, Malinalli ceases "to be 'The Tongue' to become simply a woman, silent, voiceless, a mere woman" (79). In essence, her body, land, and religion are raped. It is in this moment that Malinalli draws her strength from her pre-colonial memory. Malinalli returns to the images of interconnectedness and mediation through the memory of her ancestors, and in her tears, she understands that water is language and "is the essence of our god who brings together extremes and reconciles the irreconcilable" (79).

Malinalli cannot understand Cortés's disconnection from her body and the land, and in order to reconcile their worldviews, she asks him to join her in a purifying ritual in the bathhouse. Because Cortés suspects the Mexicas are going to attack, he hesitates to enter a closed space in the nude, but he enters in order to gain knowledge of Quetzalcóatl and his own connection to that deity (as Montezuma believed Cortés was Quetzalcóatl who did not need human sacrifices and was therefore feared by the Aztecs in power who utilized sacrifice in their religious practices). Cortés's goal is to gain information about Quetzalcóatl for his own material gain (90). He does gain knowledge in this space, but he is also portrayed as a man who cannot enter the spiritual world of Malinalli—only the physical.

His desire to subdue Malinalli proves true as her knowledge and voice becomes his in her desire to translate his words correctly: "She spoke in the name of Malinche, a nickname they had given Cortés, since he always had her by his side. Malinche in some way meant the master of Malinalli" (93). The ugliness of Malinalli's conquest is made apparent when Cortés and his soldiers do not use her translation skills, but instead prematurely enter war with the Cholulan peoples, killing over six thousand Cholutecans. Malinalli carries the weight of the dead on her shoulders, while, "For Cortés, the conquest was a struggle of good against evil, of the true gods against the false gods, of superior beings against inferior beings" (95). Archeologist Geoffrey McCafferty connects the conquest of land to religious beliefs when he writes, "The Tlaxcalteca account of the battle emphasizes the religious aspect of Cholula, relating a native belief that any assault on the temple of Quetzalcoatl would result in devastating floods that would wash away the attackers; it was only after that prediction failed to occur that the Tlaxcaltecans joined in" (351). As Cortés begins to believe that he is "saving all the Indians for the ignorance in which they lived" (Esquivel 95), Malinalli begins to question her safety and who she should serve—the Spanish or her people. This confusion occurs because two stories are colliding, blending, and clashing—that of Quetzalcóatl and that of Cortés's ideas of conquest.

Subsequently, as Malinalli's story and perspective become the focus of the novel, it raises awareness of the misuse of rhetoric for specific colonial purposes. In much the same way that Cortés twisted Malinalli's words for his needs, the Black Legend was a rhetorical strategy used to taint the reputation of the Spanish empire. This is seen when Ixtlilxochitl omitted Malinalli's importance in the conquest, and when Cortés himself failed to mention her in his letters to Spain. The absence of her mention in these letters and documents illustrates her importance in the conquest. She was not used to present Spain's accomplishments to the world, as her gender would lessen the importance of Cortés's claims. This omission allows for the conquest to be focused on the male figures in war. Cortés and Montezuma become polarized figures, erasing the voices of Malinalli and the indigenous populations of New Spain. Thus, the legend of Malinche within the Black Legend is used to illustrate the weaknesses of Spain to the Protestant English, and the weaknesses of the Indigenous to the Catholic Spanish.

The text complicates rape as not just violence, but as an outcome of a lack of interconnectedness. Toward the end, Malinalli acknowledges the role her body has taken in the conquest and the supremacy of the words she has used as a translator for Cortés. Esquivel draws on the symbols of water, tongue (masculine), and mouth (feminine) to illustrate unity. On her wedding night, Malinalli pierces her tongue with a thorn from an agave plant after a drunk Jaramillo repeatedly rapes her (158-59). Because Malinalli pierces her tongue, which is her "instrument of the mind," the expedition to conquer Hibueras fails, and Cortés's defeat is then "buried in silence" (159). But not long after, "tired of being Cortés's reflection" (177), she commits suicide and is once again connected to the land, water, and her deities. She walks into the center of the canal which was, "the Cross of Quetzalcóatl, the center of all crossroads, where the Cihauteteo, the women who had died during childbirth, who made up the entourage of the women who follow Tlazolteotl, Coatlicue, and Tonantzin, the different manifestations of the same feminine deity, appeared; and there, in the center of the Universe, Malinalli became liquid" (184-85). In this moment, the union between mouth and tongue are separated through death, but understood as necessary in legend—allowing for the continuation of the legend of La Malinche and the Black Legend. The stories continue because there is a disassociation from balance.

CONCLUSION

The Black Legend draws some of its lasting power from La Malinche as it illustrates Mexico's own need to connect itself to a European power rather than to honor indigenous history. The importance of understanding the connection between land conquest and legends attached to the land should be a part of the discussion of the Global South. As Richard Rodriguez states in "India," "Mexico's tragedy is that she has no political idea of herself as rich as her blood" (19). Thus, work must continue to fully engage in a history that disengages Malinalli and Mexico from the rhetoric of the legend of La Malinche and the Black Legend in order to change perspectives about indigeneity and the conquest of land.

WORKS CITED

Candelaria, Cordelia. "La Malinche: Feminist Prototype." *Frontiers: A Journal of Women Studies* 5.2 (Summer 1980): 1–6. Print.

Castro, Rafael G. *Chicano Folklore: a guide to the folktales, traditions, rituals and religious practices of Mexican Americans*. New York: Oxford UP, 2001. Print.

Cypess, Sandra Messinger. "'Mother' Malinche and Allegories of Gender, Ethnicity and National Identity in Mexico." *Feminism, Nation and Myth: La Malinche*. Ed. Rolando Romero and Amanda Nolacea Harris. Houston: Arte Publico P, 2005. 14–27. Print.

De Las Casas, Bartolomé. *The Devastation of the Indies: A Brief Account*. Trans. Herma Briffault. Baltimore: Johns Hopkins UP, 1974. Print.

Esquivel, Laura. *Malinche*. Trans. E. Mestre-Reed. New York: Washington Square, 2006. Print.

Gonzales, Sandra M. "Colonial Borders, Native Fences: Building Bridges between Indigenous Communities through the Decolonization of the American Landscape."

———. *Comparative Indigeneities of the Américas: Toward a Hemispheric Approach*. Ed. María Bianet Castellanos, Lourdes Gutiérrez Najera, Arturo J Aldama. Tucson: U of Arizona P, 2012. 307–20. Print.

Griffin, Eric. "Nationalism, the Black Legend, and the Revised Spanish Tragedy." *English Literary Renaissance* 39.2 (2009): 336–70. Academic Search Premier. July 2012.

Maltby, William. S. *The Black Legend in England: The Development of Anti-Spanish sentiment 1558 – 1660*. Durham, NC: Duke UP, 1971. Print.

McCafferty, Geoffrey. "The Cholula Massacre: Factional Histories and Archeology of the Spanish Conquest." *The Entangled Past: Integrating History and Archeology*. Ed. M. Boyd, J.C. Erwin, and M. Hendrickson. Academia.edu. Web. April 2013. 347–359. 15 April 2013.

Mignolo, Walter D. "Afterword: What Does the Black Legend Have to Do with Race?" *Rereading the Black Legend: The Discourses of Religious and Racial Difference in the Renaissance Empires*. Ed. Margaret Greer, Walter D. Mignolo, and Maureen Quilligan. Chicago: U of Chicago P, 2007. 312–24. Print.

Montaigne, Michel. *Essays of Michel de Montaigne*. Trans. Charles Cotton. Garden City, NY: Doubleday, 1947. Print.

Orquera, Yolanda Fabiola. "'Race' and 'Class' in the Spanish Colonies of America: A Dynamic Social Perception." *Rereading the Black Legend: The Discourses of Religious and Racial Difference in the Renaissance Empires*. Ed. Margaret Greer, Walter D. Mignolo, and Maureen Quilligan. Chicago: U of Chicago P, 2007. 167–87. Print.

Paz, Octavio. "The Sons of La Malinche." *Goddess of the Americas/La Diosa de las Americas: Writings on the Virgin of Guadalupe*. Ed. Ana Castillo. New York: Riverhead, 1996. 197–208. Print.

Powell, Phillip Wayne. *Tree of Hate: Propaganda and Prejudices Affecting United States Relations with the Hispanic World*. New York: Basic, 1971. Print.

Rodriguez, Richard. "India." *Goddess of the Americas/La Diosa de las Americas: Writings on the Virgin of Guadalupe*. Ed. Ana Castillo. New York: Riverhead, 1996. 17–24. Print.

Said, Edward. *Orientalism*, New York: Vintage, 1978. Print.

NOTES

1. Bartolomé de Las Casas writes of one catastrophic event where the Spanish burned eight hundred towns in "the kingdom of Jalisco" (78). He speaks to his Spanish audience, stating that the Indians have due cause to force the Spanish out of Jalisco, "under all the laws, natural, human, and divine" (79).

2. Translated, this means, Long live Mexico! Sons of the screwed woman.

3. La Malinche was Hernán Cortés's translator and lover. Malinche and Hernán Cortés are attributed to having had the first mestizo son in the New World, giving her the legacy of being a traitor to the Mexican people.

4. Malinche's many names include Marina, Malinche, Malintzin and "The Tongue," which maintain colonialist and patriarchal attitudes. Malinalli's birth name is returned to her in Esquivel's novel. I choose to do the same from this point on unless referencing the legend itself, offering historical accuracy and depth to this legend.

5. Ixtlilxochitl (1568-1648) was a Mexican historian. His work was very critical of the taxes the Spanish crown was imposing. He discussed the atrocities of the invaders' conquest and gave voice to the poor left in New Spain.

ELEVEN

Mapmaking, Rubbertapping: Cartography and Social Ecology in Euclides da Cunha's *The Amazon*

Land Without History

Aarti S. Madan

In his allegorical "Tale of the Giants," Giambattista Vico locates the origins of Western civilization in the nebulous boundary between the habitat and the forest (*New Science*, 1774). Civilization, he explains, began when prehistoric giants became inhabitants of the clearing. Since only open spaces allowed views of the heavens, the giants necessarily cut trees and, from there, established their belief systems. But to be civilized, following Vico, means more than to clear; it means to differentiate between the clearing and the forest, to pay due heed to the very act of delineation and demarcation, and, in the end, to respect the otherness of the forest. The giants' descendants—indeed, Western civilization—nevertheless explode the limits of the forest, thereby erasing all traces of their own culture, origin, and identity.

In this article, I aim to explore the ways in which Euclides da Cunha dramatizes the cartographic encounter of the imperialist West with the otherness of the Brazilian Amazon in his collection of essays *The Amazon: Land Without History* (1909). A Vichean sense of ecology informs da Cunha's literary work, yet its relationship to his geographic oeuvre remains insufficiently understood. Though Rex Nielson, in his 2013 chronology of ecological writing in Brazil, situates the compilation as "an example of autochthonous Brazilian ecocriticism," his genealogical effort is largely removed from da Cunha's cartographic practice (4).[1] And while Lúcia Sá (2006) does indeed hold imperialism responsible for the Amazon's environmental degradation, she finds Brazil and, more pointedly, da Cunha's own "imperialist nostalgia" most culpable for eliminating cultures and damaging the environment (xiv). Without reifying da Cunha or losing sight of his neocolonial tendencies, I propose an alternative model of understanding da Cunha's relationship to Brazil's ecology wherein he transculturates mapmaking, thereby creating a model of sustainability for Latin America. His project can be read as a sort of conservationist cartography, distinct from the commercial exploits that facilitated the region's rubbertapping epoch. Such a reading would allow for a new ecocritical understanding of historical projects that, on their face, might seem unimportant, or even counter, to the sensibility of environmental writing but that illustrate the unique locus of the Global South.

For da Cunha, there appears to be a direct correlation between settlement of the land and its preservation—in other words, clearing and populating the giants' allegorical forest while always valuing its alterity, indeed, always nurturing civilization's sylvan roots. To unpack da Cunha's ecologically minded project of statecraft, I look to historian Raymond Craib's distinction between *fugitive landscapes* and *state fixations*. If the former identifies "lands characterized by multiple political jurisdictions and use rights, indeterminate borders and inconsistent place-names, and highly contextualized systems of tenure and property" (Craib 12), then da Cunha seeks the latter for the Brazilian Amazon: a "state fixation" committed to permanence through "the inscription of lines, points, plots, and place-names" that "would give space a stable signification, permitting it to be more effectively appropriated, transformed, and regulated" (Craib 8). More to the point, I argue that da Cunha looks to overcome the destruction of the Amazon's transient rubber industry—most visibly characterized by the deforestation of the Peruvian *caucheros*—by naming and demarcating the territory of the Brazilian *seringueiros*, tappers known for their particular style of extracting rubber without felling trees.

Understandably, such a nationalistic and cartographic project of enclosure and delineation does not immediately evince ecocentric designs and might even concern an ecocritic like Glen A. Love, who in *Practical Ecocriticism* (2003) appears to begrudge narratives that extol anthropocentric matters. Yet I will show that da Cunha does enact a social ecology—one that is self-conflicted and perhaps even self-contradictory, but that is therefore emblematic of the Global South: put plainly, if Brazil does not "fix"—*fijar*—its own space and, subsequently, settle and extract its resources responsibly, imperial forces will encroach upon the fugitive landscape and do just that, albeit *irresponsibly*.[2] By untangling the political, racial, and economic dramas that develop in da Cunha's Amazon, then, I will illuminate the ways in which his collection—despite its many contradictions—exhibits a deep respect for the otherness of the forest and critiques global economic forces that find sole merit in the Amazon's use-value rather than in its human and non-human occupants.[3] Ultimately, my reading of da Cunha demonstrates that traditional ecocritical theory may not suffice for understanding the fault lines between society and ecology in the Global South, where a history of creative destruction and murky borders necessitate unconventional ecological solutions.

To develop these points, what follows is divided into three sections: the first explores the relations between geopolitics and environmental change in the Americas as manifested first through Alexander von Humboldt and later, more specifically, through da Cunha's mapmaking and discursive ordering of Brazilian space; the second analyzes the mutually informative relationship of rubbertapping and cartography in *The Amazon: Land Without History*, a collection committed to state fixations by inscribing the land with both place names and settlement of rubber territories; the third and final section focuses on da Cunha's transcontinental alignments to showcase the ways in which his ecological limitations are, in fact, illustrative of early environmental thought emerging from the Global South.

A. MAPMAKING AND (HUMBOLDTIAN) EXTRACTION

In a 1905 letter dispatched from the port city of Manaus, Euclides da Cunha offers a stinging assessment of the Amazon, "the strange sadness prompted by this expansive land, marvelous and boring, without any relief where the eye might rest; and, above all, the tumult, the indescribable disorder" (*Correspondência* 255).[4] His cartographic eye unable to find rest, it obsesses over the unordered space, the vast expanse, the muddle and the mayhem. Land surveyor, military engineer, and legendary author of Brazil's canonical narrative *Os Sertões* (1902), da Cunha ultimately coalesces his technical and discursive skills to order the Amazon—that uncharted enigma heretofore unknown, indeed, unwritten—and make of the "expansive land" quite the opposite: charted, known, written, available for entry into the annals of Brazilian and, more importantly, universal history. Yet he names his writings, a collection of essays published posthumously in 1909, *À Margem da Historia*—in English, *The Amazon: Land Without History*. This title is the first of many contradictory elements informing the text's genesis.

The tripartite collection emerges from da Cunha's experiences in the Brazilian Amazon over the course of some twenty months (1904–06) in which he served as head of the Commission for the Reconnaissance of the Upper Purus River, a tributary of the Amazon. The Baron of Rio Branco, sitting Foreign Minister, purportedly appointed the expedition to litigate a series of border disputes between Peru and Brazil. From 1893–1909—over a mere sixteen years—Rio Branco had successfully redefined the borders of Brazil by inching into large tracts that had been demarcated by the 1494 Treaty of Tordesillas but disputed thereafter. Through this diplomacy, the newly acquired land—a little from Bolivia, a bit from Uruguay and Peru, still more from Argentina—increased the Brazilian territory by an area the size of France. Rio Branco's gradually expanding frontier, along with his sudden expansion of the Brazilian navy, caused anxiety across the region, prompting further territorial dispute. Although the conquest relied on words rather than wars, Brazil's expansion was understandably litigious and often necessitated external arbitration. Two years prior to *The Amazon: Land Without History*, for example, da Cunha had published his assessment of a land dispute between Perú and Bolivia, translated into Spanish as *La cuestión de límites entre Bolivia y el Perú* (1907) and arbitrated by the Argentine government. The Brazilian government that commissioned da Cunha—led by Rio Branco—could not but defend Bolivian rights in the case, for if Peru were to win, then Brazil would lose territory in the northwestern state of Acre.

By recounting these episodes of expansion-related arbitration—particularly in relation to the rubber-laden state of Acre—I aim to situate da Cunha within a moment characterized by heated exchanges regarding border demarcation, which is the task that initially placed him in Amazonia and prompted the essays under examination. Brazil had been on the winning side of territorial feuds since the time of the *bandeirantes*—those seventeenth-and eighteenth-century slave-hunting groups who, over the course of two violent centuries, expanded Portuguese territory with the threat of harquebus. But with Misiones (the site of Iguazú Falls) only marginally won in 1895 and Acre nearly lost in 1907, the tenuous nature of imaginary lines scratched into dirt quickly became all too clear; only textual delineations could create the permanence necessary for a state fixation.

Da Cunha's discursive practice thus proved a direct complement to Rio Branco's diplomatic agenda in the Amazon. The Baron sought to prevent encroachment upon Brazil's *fronteiras mortas*—"uninhabited" and therefore ill-defined territories—while also ensuring economic control of the day's most lucrative natural resource, one that literally grew on trees: rubber. Until the turn of the twentieth century Brazil and the countries that share the Amazon basin (i.e., Bolivia, Venezuela, and Peru), were the only exporters of natural rubber. In fact, during the late-nineteenth century, Brazil sold almost ninety percent of the total rubber commercialized in the world. As Susanna Hecht and Alexander Cockburn make clear, however, "Brazil's Amazon scramble was a nineteenth-century form of nation building replete with explorers, freebooters, speculators, fraudulent maps, spies, plenipotentiaries, competing and contradictory court judgments, and romantic revolutionaries" (vii). It was, in sum, anything but neat. And cartography was at its core.

As these multiple murky borders and contested territories show, an important driver of environmental change in Amazonia was—and is—geopolitics, a term I use in the more classic state-centered sense (following Ratzel and Mackinder) to describe territories wrestling for control of space. The Amazon Basin's international boundaries were largely defined according to the scramble for resources before and during the turn of the Rubber Boom. With borderlines in flux, states became increasingly concerned as they attempted to establish consistent national identities throughout their demarcated territories. The written word became just the vehicle to not only define a singular Brazilian identity but also to consolidate its geographic territory.

Da Cunha spearheaded this initiative within two distinct spaces: first, the drought-ridden backlands (*sertão*) of the Northeast, and later, the Amazonian jungle of the Northwest. While the *sertão* dominated a tale ostensibly about the Canudos War (*Os Sertões*), the Amazon played the lead in a narrative about rubber (*À Margem da História*). Both texts reveal da Cunha's misgivings at the thought of non-Brazilians traversing and charting Brazilian lands, an integral component of what he calls "modern expansionist imperialism," which he holds responsible for a "campaign against the climate" in addition to the *hygienification* of the Global South (33–34). In the essay "This Accursed Climate," he explains:

> A patent of nobility exists in modern expansionist imperialism, which serves to absolve it of the greatest crimes, namely, that its brilliant generals are transformed into anonymous scouting parties of physicians and engineers; the greatest battles merely recall the greater campaign, the campaign against the climate; and domination of the inferior races is the beginning of the redemption of territories in a magnificent sweep that hold out to all parts of the globe the marvelous undertaking of the hygienification of the land, from Tonkin to India, to Egypt, to Tunisia, to the Sudan, to the island of Cuba and to the Philippines. (33–34)

This passage exposes da Cunha's suspicion of "anonymous scouting parties" treading upon and dominating both human subject and environment in service of their expansionist endeavors, this across the Global South. His concern was not based on foresight but rather history, which can be summed up in one Prussian name: Alexander von Humboldt, perhaps the most renowned leader of a "scouting party" in the Americas.

Any ecological interrogation of North-South relations in Latin America would be remiss to ignore Humboldt and his multiple volumes describing the Americas. Like Marco Polo's exultant tales of the Orient and Columbus's fabulous accounts of naked natives, lush vegetation, and abundant gold, so Humboldt's scientific travelogues and cartographic endeavors literally drew a geography that would captivate and create in both the Old and the New Worlds, both in his times and in our present.

Humboldt landed upon America's soils armed not with weapons and shields but rather classifying devices that included everything from barometers to compasses, from charting instruments to astrolabes, and, most importantly, from feather pens to papers. These pens and their subjective recordings would, according to Simón Bolívar, change the face of America: Humboldt, declared Bolívar in an 1821 letter, "has extracted America from ignorance and, with his pen, has painted her as beautiful as her very nature" (in Humboldt, *Cartas Americanas* 266).[5]

His positive role in Latin American independence struggles aside, with Humboldt surfaces the question of commercial cartography. If, for Marx, in the bourgeois era the world appeared as a spatially diversified bundle of natural resources waiting to be discovered and exploited and extracted, then commercial maps reflect this trend. The backdrop for Humboldt's work is mercantile. He constructs a systematic description of the earth's surface as a repository of use-values. Humboldt—the primordial Eurocentric man—views America as *materia prima*, as space to be ordered in the service of modernity. In *Views of Nature* (1807), he deems the Amazon jungle a "primeval forest," designated as such due to its impenetrability and its autopoetic system that makes it continuously feed on its surroundings, which then turn into wood; prehistoric and precultural, it exists only as a radical other of humanity in general and of Western civilization in particular. A space devoid of human beings, then, the Amazon is valuable only for its material profits. In a different context and after some years, Humboldt declares quite markedly in *Political Essay on the Kingdom of New Spain* (1811) that "a colony has for ages been useful to the parent state only in so far as it supplied a great number of raw materials" (185).

And supply they did: Humboldt's maps ignited the South American guano boom and mining rush, both of which filled the coffers of countless European investors but decimated local populations, lands, and economies. With his complicity in colonial power structures evident, Humboldt's charting and naming of American lands reveal themselves as colonial continuations of "the discourse of enclosure," which begins in seventeenth-century England as communal properties become privatized. This turn, explains Robert Marzec, transforms the *inhabitant*—a human coexisting within the environment and the community—into an individual subject removed from the natural world: a worker, a tenant, a landlord (424). At this crux, the inhabitant exchanges precapitalist attention to habitat in favor of capitalist extraction for survival, which is to say, another manifestation of Vico's new variant of barbarism.

Euclides da Cunha writes from precisely this intersection. Caught between worry for the Amazonian environment and an incongruous interest in sustaining Brazil's capitalist economy, he composes a collection of essays that are just as fragmented as his perception of the land. From the very first lines of the opening essay—"General Impressions," in which

Alexander von Humboldt's name appears three times in just the first six pages—da Cunha begins enumerating and correcting the Europeans who had previously described the region.[6] Welsh naturalist Alfred Russel Wallace, whose awe at the sight of the Amazon's waters only stymies a disenchanted da Cunha, receives the first round of skepticism. "Rather than admiration and enthusiasm," da Cunha admonishes, "what usually overcomes someone beholding the Amazon at the point where the Tajapuru's vibrant confusion flows full into the great river is a great sense of disillusionment" (*The Amazon: Land Without History* 3). He continues:

> To be sure, the sheer volume of water is unmatched and therefore capable of inducing that wonderment of which Wallace speaks. But since, from early on in life, each of us had drawn an ideal Amazonia in our minds thanks to the remarkably lyrical pages left us by the countless travelers, from Humboldt down to today, who have contemplated the prodigious hylean rain forest with almost religious awe, we experience a common psychological reaction when we come face to face with the real Amazon: we see it as somehow lacking with respect to the subjective image we have long held of it. (3)

All his criticisms aside, for da Cunha, Amazonia is authentically Brazilian—and therefore deserves to be written by an authentic Brazilian—so much so that his on-site correspondence fluctuates between deploring the vast expanse, the humidity, the danger (he at times expresses great concern for his life) and praising the Amazonian port of Manaus over and above Rio de Janeiro as a fitting window into Brazil.[7] In a January 1905 letter to Afonso Arinos, he first describes Manaus as the "noisy, large, unorganized, monstrous and opulent capital of the *seringueiros*," where fevers rack his body, yet he closes by praising its cosmopolitanism and wondering why "this, our Brazil," stills exists as periphery on both local and global levels. "Really," he muses, "I am more and more convinced that the deplorable Ouvidor Street is perhaps the worst prism from which to view our land" (*Correspondência* 251).[8]

Da Cunha thus reveals himself as profoundly affected by the Amazon, seeing that, on some level, it is akin to and a better reflection of Brazilian otherness. It should not be represented as comprehensible or definable or containable because doing so—in the footsteps of Eurocentrics like Alfred Russel Wallace, Alexander von Humboldt, and Henry Walter Bates—is to define or to contain its power. But it must be written, and poetically at that, for it is "humid and hot, made for the fibers of palm trees and the nerves of poets" (*Correspondência* 257).[9]

Why such concern with authenticity, with the "real Amazon"? Why the need to rewrite Wallace and Humboldt and to disagree with Bates, whose token expression regarding the Amazon's "glorious clime" bemuses and confuses da Cunha enough to reference in three separate letters (*Correspondência* 250; 252; 255)? I sense here a continuation of da

Cunha's correctives in *Os Sertões*, wherein he reclaims both lands and letters by reframing the works of Europeans who had previously traversed Brazilian territory.[10] Authenticity, here, is tantamount to conservation: in other words, only a Brazilian is invested in preserving Brazil for posterity. Da Cunha's task in composing the essays of *The Amazon: Land Without History* is thus twofold. First, he desires to *write the earth*—the etymology of geography—to enter Amazonia not only into Brazilian but also universal history. This is his discursive practice, one embedded with a certain didactic impulse.

Second, and on the technical level, da Cunha's task as a cartographer and land surveyor is to ensure that the land will be mapped and therefore settled. He transculturates—indeed, Southernizes—the "discourse of enclosure" as model of sustainability: it emerges as a sort of conservationist cartography, separate and distinct from the commercial and exploitative. For da Cunha, there appears to be a direct connection between settlement of the land and preserving it, indeed, historicizing it by means of population. He distinguishes between "cultivated landscapes," replete with "the artistic touches that human labor adds" and the opposite, which punctuate the Amazon:

> Here, such is simply not the case. [. . .] [O]ne gets the sense that its impenetrability would go on forever did it not abound in precious products directly acquirable without the need for the constancy of cultivation. The people who live in the Amazon work it in the wild. They do not cultivate it, thereby making it attractive; they merely tame it. (32)

Ever the aesthete, da Cunha finds little art or beauty in the act of taming. To extract the precious products is only to temporarily tame the land; creative destruction is thus not so creative. To truly—and permanently—cultivate and create upon it, on the other hand, requires overcoming the Amazon's impenetrability by charting and documenting its contours in ways that stand the test of time. In this sense, then, mapping is a form of historicizing; history becomes absolutely inextricable from geography and, with that, cartography.

In the last essay of his compilation, "Transacreana," da Cunha explains that "the map of Amazonia is a diagram of its initial settlement. The history of the new region, rather than being written, must be drawn. It isn't read, it is seen" (77). The map is thus both instrumental and iconographic, a representation not only of past but a confirmation of present and premonition of future. Uncharted land, on the contrary, constantly evolves, much to the detriment of civilization, and therefore requires settlement. Of course, da Cunha's aspirations to order Brazilian space must be contextualized within his own *local* Eurocentricity, necessarily different from the European and North American writers he imbibed daily but equally committed to ideas of settlement solidified during the colony. Being "settled," here, is tantamount to settling down, which would lead the Brazilians to cultivate the land and, in turn, their own modernization.

B. RUBBERTAPPING: *SERINGUEIROS* AND SETTLEMENT

The Amazon: Land Without History is not only a descriptive text of the Amazonian river region but also a prescriptive account of rubber's role in the economics of nation-state consolidation. Euclides da Cunha was in no way averse to extractive commerce, particularly of a material so essential to Brazil's growth. Like his European predecessors, he was convinced that the *fronteiras mortas* of the Amazon were mere nature, "vast, unpopulated, almost known" and therefore not only in need of population but also exploitation (36).

Yet those lands were long inhabited by the Amazon Basin's indigenous communities, who looked to natural rubber for a variety of uses including rubber balls, waterproof clothing, and torches. They were also the first generation of (extraordinarily abused) early-nineteenth-century-boom rubber tappers (Grandin 30). Their Dickensian plight canonized through Latin American letters (e.g., Alfredo Rangel's *Inferno Verde* [1908], Jose Eustasio Rivera's *La Vorágine* [1924], Rómulo Gallegos's *Canaima* [1935], Mario Vargas Llosa's *La casa verde* [1966], to name just a few examples), the tappers suffered horrifically during the so-called "scramble for the Amazon" and, subsequently, after the collapse of the rubber industry.[11] But during the height of the Boom—set off, incidentally, by none other than the bicycle—the tappers had plenty of work, indentured servitude though it may have been. Europeans were voraciously using rubber for waterproof shoes and later, after Charles Goodyear's 1839 improvements of vulcanization, for uses "ranging from sprockets, belts, hoses, and tiles to suspenders, shoes and raincoats" (Weinstein 8). The Boom culminated with Henry Ford's ill-conceived plan to transform part of Brazil's Amazon Basin into a rubber plantation, a fascinating tale fictionalized in Eduardo Sguiglia's *Fordlandia* (1997) and marvelously documented in Greg Grandin's more recent historical study by the same name (2009).

Rubber functions as the most contradictory symbol in *The Amazon: Land Without History*. On its face, it reveals the narrative's western logic wherein all interactions between the human and nonhuman world condense into relations of profit. With varying degrees of condemnation, da Cunha details a floundering economy whose powers have been extinguished in tandem with the slow collapse of the Amazonian environment. Rubber, in the end, enacts an epistemology of separation wherein (white) humans exist separate from their physical environment. Yet da Cunha dedicates a disproportionately small space to addressing the human component in this and other texts; this neglect has resulted in the brunt of past and recent criticism, most notably Lúcia Sá's aforementioned Introduction and Adriana Michèle Campos Johnson's recent study *Sentencing Canudos: Subalternity in the Backlands of Brazil* (2010).

While such criticisms are, to a large degree, valid, they overlook the value of da Cunha's shortcomings and what those failures might point to: a social ecology that is fitted to the specificities of a recently decolonized, unstable, and therefore vulnerable territory. If part of da Cunha's task is to fortress Brazil against the Global North, an equal part is dedicated to declaring his nation's regional superiority, particularly in the context of rubber. These declarations, to be sure, extend to Brazilian rubber tappers. In his penultimate essay, titled "Brazilians" in quotation marks, da Cunha lists a series of comparisons between Peruvians and Brazilians, each of which concludes with an unsurprising conclusion about who is superior. All the while, he laments the deplorable conditions of the rubber tappers, their monotonous life, their infirmity, their debt and indentured servitude. Da Cunha pays special attention to differentiating between the Peruvian *cauchero*, who collects the rubber from the *Castilloa* tree, as opposed to the Brazilian *seringueiro*, who collects from the *Hevea*. The Peruvian must fell the tree to access the rubber, whereas the Brazilian taps via small cuts into the bark of the tree, which survives and can be sustainably harvested in future years.

Here we arrive at the critical heart of da Cunha's social ecology. He explains:

> The exploitation of caucho as the Peruvians practice it, with its felling of the trees and the constant movement in search of undiscovered stands of *Castilloa* in an endless professional nomadism, leads them to practice all manner of abuse in the inevitable confrontations with the natives, and thus brings with it the systematic disruption of society. The cauchero, eternal seeker of new locations, has no relationship to the land. In his primitive activity, he develops prioritarily the attributes of cleverness, agility, and strength. It involves, in sum, a barbarous individualism. There is a lamentable involution in a man perpetually away from society wandering from river to river, from thicket to thicket, ever in search of virgin forest in which to hide or take refuge, like a fugitive from civilization. (71)

With his "systematic disruption of society," his "primitive activity," and "barbarous individualism," the Peruvian *cauchero* does not respect the jungle's otherness. Rather, "he has no relationship to the land," which he deforests with little reflection upon the ways it determines his identity, his livelihood, his salvation. If the Brazilian *seringueiro* is Vico's prehistoric giant, then the Peruvian *cauchero* might well be his radical descendant, little more than a "fugitive from civilization," a "man perpetually away from society" and therefore fitting for the Amazon's fugitive landscape—

the antithesis of a harmonious inhabitant of a pre-enclosure, precapitalist land.

But though the Brazilian *seringueiros* respect the forest's otherness and value its longevity, they have, in da Cunha's estimation, insufficiently populated the territory. In the essay "This Accursed Climate," for example, da Cunha establishes the dearth of inhabitants in Acre, noting, "Above all else is abandonment" (39). He continues:

> Even in the Acre proper, where the greater density of rubber trees permits the opening of sixteen sections, or 'paths,' per square league, that area—capable, according to the current calculation of the agricultural unit, of supporting the families of fifty subsistence farmers—requires the labor of only eight *seringueiros* spread out over the terrain, rarely catching sight of one another. [. . .] And that latifundium, which could be populated with up to three thousand active inhabitants, holds no more than the invisible population of one hundred exaggeratedly dispersed workers. (39)

The numbers say it all: 100 thin implies a fugitive landscape, whereas 3000 strong signals a state fixation: Brazil is there to stay, quite the opposite of what Brazilian General Golbery do Couto e Silva would many years later call a *fronteira morta,* "a dead border, barren, unprotected, and forgotten," nothing more than "an invitation for invasion" (Salisbury 52).

Beyond the threat of encroachment, abandonment also breeds further barbarism. In the essay "The Caucheros," da Cunha recalls a particularly grisly July 1905 visit to a *puesto cauchero* on the Purus River from which the lasting image sealed upon his mind's eye is that of "a naked, atrociously mutilated body lying in a clearing amid the *frecheira* trees on the left bank of the river" (54). While he expresses dismay at how little his commission learns regarding the cadaver, he attributes the apathetic gloss-over to abandonment; murder, after all, is "a highly trivial, virtually meaningless matter in a place frequented by people who pass through but do not settle. They move on, leaving it even sadder with the remains of abandoned *estancias*" (54). Da Cunha seems less concerned with the human remains and more so with the spatial destruction, with the lack of civilization at this arboreal edge where a new incarnation of barbarism has taken over the abandoned forest.

As proxy of the civilized state, then, da Cunha aims to establish some sort of permanence. The act of naming and consistency are integral to such an effort. Yet he explains that in Amazonia, names, too, are transitory:

> In the land without history the first facts are written, scattered and fragmentary, in place names. First comes the initial, torturous phase of adaptation, which evokes sadness, martyrdom, even cries of despair or calls for help. The travelers read on the large signboards on the sides of the houses turned out to the river: "God Save Us," "Homesickness,"

> "Saint John of the Misery," "Hidden," "Hell" . . . Then the strong rebirth of hope and overflowing good humor of a people redeemed: "Good Beginning," "New Enchantment," "I Want to See!" "Liberty," "Concord," "Paradise." (41)

But why such emphasis on onomastic consistency, something so benign at face-value? For da Cunha, place-names preserve history by inscribing a genealogy upon the land. In the cosmovision of the day, reflected by da Cunha's Mexican contemporary Antonio Peñafiel, place names preserved tradition in places where "history has completely disappeared" (qtd. in Craib 44). Land without history is, in fact, land without *geography*—land where the earth remains to be written. Da Cunha looks to overcome just that limitation, thus provincializing Brazilian geography and allowing his nation-state to enter modernity.

So whether settlement through sheer numbers or through place-name permanence, da Cunha seeks fixity; further, he deems the federal government responsible for such measures. To his mind, the threat of neighboring states' invasion and encroachment diminishes in the face of *fronteiras vivas*, signs of civilization further buttressed by the ultimate symbol of progress: *Transacreana*. Sealing his compilation with the thunderous rumble of modernity, the last essay "Transacreana" accounts for da Cunha's proposed rail-line, which "would tie together the three administrative seats" through "a length of 726 kilometers, or exactly 110 leagues, which can be covered by horse in fewer than twelve days" (82–83). Da Cunha suggests the rail-line as an auxiliary to the rivers, which, at the time of his writing, controlled the rubber economies. If the tapping of rubber trees "derives merely from the courses of the rivers," then "the new rail line will, naturally, represent a 'path' providing entry for new workers" (87). He insists that "[t]he intervention of the federal government is urgently needed to perform the basic task of linking together the several discrete initiatives and giving incentive to the project" (82).

Steel cutting through the space of nature: nothing more permanent than this could suffice in da Cunha's imaginary of the developed Amazon. Of course, such aggression upon the forest tests the parameters of environmentalism, as does reducing the chaos of the forest to an ordered grid. But easier transport of rubber ensures economic growth, while both railways and maps consolidate the nation's space in ways hindered by its very geography. The two-dimensional writing of maps motivates, allows, and even determines the demarcations etched through the jungle's three-dimensionality. And, as a representative of the state, da Cunha takes due care to underscore his cartographic expertise, stating that "[i]t should be observed in passing that this project is not one drawn up according to the arbitrary lines to which 'map explorers' are accustomed": "It rests instead on practical knowledge," he insists, while adding the qualifier, "absent azimuths or aneroid readings, but concrete nonetheless and therefore definitive" (82).

In a true postcolonial "search for authenticity" (Said 226), da Cunha looks to create a unique Brazilian cartography, one perhaps devoid of European luxuries like azimuths and aneroid readings but nevertheless grounded on authentic and "practical knowledge." If the nation-state and its most chaotic spaces can be ordered with grid-lines and rail-lines—if the experienced natural setting can be wrangled into submission—then da Cunha can safeguard Brazilian space from global economic forces. Ultimately, then, the dual narratives of map-making and rubber-tapping inform one another in *The Amazon: Land Without History*, for both engage questions of national and environmental security. By emphasizing state fixations—through settlements, through place names, through trees that are permanent rather than felled—da Cunha seeks stable production rather than nomadic destruction. Key here is the *Transacreana*, of which he writes "[t]he economic value of the proposed project is incalculable" (86). And with this last statement da Cunha illuminates his model for modernization: surmounting the previously insurmountable, both in terms of nature and nation. Each of these facets of progress undergirds the state's protective—and proactive—measures to ensure its steady growth, both in terms of a capitalist economy and expanding geography.

C. SOCIAL ECOLOGY AND THE GLOBAL SOUTH

Euclides da Cunha's cartography, to his mind, is authentically Brazilian. It is concrete and definitive, unlike the arbitrary lines of "'map explorers,'" to which he raises two eyebrows (or quotation marks) of suspicion (82). By reducing the jungle to a written document rather than an experienced natural setting, da Cunha can ensure that the space of Brazilian nature is protected, even preserved, from global economic forces. It is in this sense that he manifests an ecological sensibility that is fitting for the Brazilian context while perhaps wholly un-ecological in spheres never before colonized.

In his 1959 manifesto *New World in the Tropics: The Culture of Modern Brazil*, Gilberto Freyre embarks on a dogged mission that is, in some ways, a continuation of da Cunha's: to demystify Brazil for non-Brazilians (in his case, English-language readers); to enter Brazil into universal history; to elucidate that "[t]he tropics and Brazilian civilization seem to have a secret alliance against all possible enemies" (36). Freyre patiently describes Brazil's "intimate contact, and even fusion, of culture and environment" while refuting the notion that "European civilization should be preserved in Brazil just as it is kept or preserved—as if in ice—in boreal Europe" (24).

Much the same might be said about Euclides da Cunha's unique brand of social ecology. It is adapted to the warm Brazilian climes, wherein his style of essay exemplifies—in Freyre's words—"the most original expression of Brazilian literature. For it combines philosophical and social preoccupation with the artistic and literary, becoming therefore more than *belles-lettres* . . . " (14). By merging aesthetics and politics, the essays that compose *The Amazon: Land Without History* reveal a sort of ecological thinking not just fitted to Brazil but arising from its specific context of exploitation.

Yet da Cunha's collection shows us that such a social ecology is not just local, emerging from and therefore limited to the Amazon. Rather, he continually reminds us of Brazil's position within the Global South and, as such, in relation to the Global North. His essays thus repeatedly reference those comrades in arms, indeed, all products of "modern expansionist imperialism" (33). In the essay "General Impressions," for example, da Cunha identifies the Amazon as something of a consolation prize, explaining that "[t]he wild Amazonian region has always had the gift of impressing far-off civilization," so much so that "the distant metropole outdid itself in efforts to open up this land that above all others would compensate it for the lost, prodigious India" (11). Later, in "Brazilians," he highlights colonialism's transcontinental riches from *calisaya* quinine, which made its way from Peru to India courtesy of the British government's commissioned geographer, Sir Clements Markham: "He was so successful in transplanting to India that element of Peruvian fortune that by 1862 more than 4 million trees, with the incredible production of 370 tons, launched from Darjeeling a competition that triumphed on the first assault" (69).[12] To further cement the geographical similarities between the two poles, he draws alignments between *Transacreana* and the successful *Great Indian Peninsula Railway*, created by the Bombay Corp of Engineers in 1849 (83).

As scientist and engineer, da Cunha sees the value in overcoming the nation's geographic complexities to thereby boost the economy. Yet, at the same time, he is an artist dedicated to cultivating natural space in ways that fix the state into both local and global archives—be it through maps, economies, railroads, place names, or trees whose rings reveal not just natural age but also national longevity. By aligning the divergent poles of colonization's toils, then, da Cunha reveals concomitant distaste *and* marvel, suspicion *and* respect. He wants Brazilian engineering to reach the standards already set by the British in India, this so that Brazil can control Brazilian space in ways perhaps extractive and productive but nevertheless committed to conservation. While such violence to the forest pushes at the limits of any ecological sensibility, da Cunha aims to concretize the social institutions of Brazil by mapping—by grafting a formula onto the uneven contours of the forest and thereby inscribing the ordered grids of science and technology onto the face of nature.

In some ways, da Cunha exhibits a matter-of-fact and even cynical acceptance of the Global South's realities: if the Brazilian government fails to order, settle, and modernize the Brazilian state, other forces will do just that—and at a higher cost to both human subject and environment. Couched within his ecological limitations, we see a Vichean sense of ecology wherein the giants develop nature but stop short of destroying it. If, following Vico, religious and social institutions, memory and human history, originate in the clearing, then Euclides da Cunha exhibits a desire for that clearing to be carved from the Amazon by Brazilian peoples rather than by global economic forces, thereby confirming his nation's sylvan roots and, with that, its secret alliance against all possible enemies.

WORKS CITED

Craib, Raymond B. *Cartographic Mexico: A History of State Fixations and Fugitive Landscapes*. Durham, NC: Duke UP, 2004. Print.

da Cunha, Euclides. *Correspondência de Euclides da Cunha*. Eds. Walnice Nogueira Galvão and Oswaldo Galotti. São Paulo: Editora da Universidade de São Paulo, 1997. Print.

———. *The Amazon: Land Without History*. Trans. Ronald Sousa. Oxford, UK: Oxford UP, 2006. Print.

Freyre, Gilberto. *New World in the Tropics: The Culture of Modern Brazil*. New York: Knopf, 1959. Print.

Grandin, Greg. *Fordlandia: The Rise and Fall of Henry Ford's Forgotten Jungle City*. New York: Metropolitan Books, 2009. Print.

Hecht, Susanna, and Alexander Cockburn. *The Fate of the Forest: Developers, Destroyers, and Defenders of the Amazon*. Chicago: U of Chicago P, 2011. Print.

Humboldt, Alexander von. *Political Essay on the Kingdom of New Spain*. Trans. John Black. Ed. Mary Maples Dunn. New York: Knopf, 1972. Print.

———. *Cartas Americanas*. Trans. Marta Traba. Caracas: Biblioteca Ayacucho, 1980. Print.

Love, Glen. *Practical Ecocriticism: Literature, Biology, and the Environment*. Charlottesville: U of Virginia P, 2003. Print.

Nielson, Rex P. "Ecocritical Thought in Euclides da Cunha's Correspondence and Writings about the Amazon." *Hispanic Issues On-Line* (forthcoming). Web.

Marzec, Robert P. "Speaking Before the Environment: Modern Fiction and the Ecological." *Modern Fiction Studies* 55.3 (2009): 419–442. Print.

Sá, Lúcia. "Voicing Brazilian Imperialism: Euclides da Cunha and the Amazon." Introduction to *The Amazon: Land Without History*. Oxford, UK: Oxford UP, 2006: xi-xxiii. Print.

Said, Edward. *Culture and Imperialism*. New York: Vintage Books, 1994.

Salisbury, David S., L. Alejandra Antelo Gutiérrez, et al. "*Fronteras Vivas* or Dead Ends? The Impact of Military Settlement Projects in the Amazon Borderlands." *Journal of Latin American Geography* 9.2 (2010): 49–71. Print.

Vico, Giambattista. *New Science*. Trans. David Marsh. London: Penguin, 2001. Print.

Weinstein, Barbara. *The Amazon Rubber Boom 1850–1920*. Stanford, CA: Stanford UP, 1983. Print.

NOTES

1. See Rex Nielson, "Ecocritical Thought in Euclides da Cunha's Correspondence and Writings about the Amazon," forthcoming in *Hispanic Issues On Line*. Another recent ecocritical analysis of da Cunha appears in Beatriz Rivera-Barnes & Jerry Hoeg's *Reading and Writing the Latin American Landscape* (Palgrave Macmillan, 2009), which examines the dialogue between literature and ecology in da Cunha's Os Sertões but, similarly, without considering the relations of cartography to environmental collapse. On the other side of the token, Mark D. Anderson recently published a piece titled "Treacherous Waters: Shipwrecked Landscapes and the Possibilities for Nationalistic Emplacement in Brazilian Representations of the Amazon," which examines the relations between the Amazon's rivers and a sense of Brazilian geographic nationalism in the works of Euclides da Cunha and Alberto Rangel (Hispanic Issues On Line 12 [2013]). Though a cogent spatial analysis grounded in the theory of biosemiotics, Anderson's piece is concerned less with Amazonian ecology and more with nationalistic symbolism and scientific imaginary.

2. I use "social ecology" as Murray Bookchin would—that is, to suggest that the roots of ecological crisis are firmly planted in relations of hierarchy and domination.

3. For a recent Latin Americanist study that thoughtfully takes upon contemporary anthropocentric novels through an ecocritical lens, see Laura Barbas-Rhoden's *Ecological Imaginations in Latin American Fiction* (Gainesville: UP of Florida, 2011). Like Barbas-Rhoden, I agree that ecocritics must read Latin American texts with difference in mind, always already considering the ways that "texts of ecological imagination use a rhetoric of nature to expose and critique human power structures" (2). By examining nonfictional narrative from the turn of the century, however, I extend her argument both generically and temporally, this while continuing to explore how da Cunha "offers warnings about modernization and highlights paths of resistance to it" (2).

4. . . . "a estranha tristeza que nos causa esta terra amplíssima, maravilhosa e chata, sem um relevo onde o olhar descanse; e, principalmente, o tumulto, a desordem indescritível . . ." (*Correspondência* 255). All in-text translations are mine, unless indicated otherwise.

5. . . . ha arrancado [América] de la ignorancia y con su pluma la ha pintado tan bella como su propia naturaleza" (in Humboldt, *Cartas Americanas*, 266).

6. Da Cunha finds himself perturbed by the varied and inconsistent conclusions drawn by his scientific predecessors: "Vê-se bem como variam os pareceres sobre assuntos tão simples . . ." (*Correspondência* 253). [It is seen as well how opinions on issues so simple vary . . .]

7. Writing prior to his departure for the upper Purus River, da Cunha ominously portends that the Amazon will devour him: "Estou a dois passos do deserto e nas vésperas de uma viagem, inçada de tropeços, dessas em que a gente leva carta de prego para o Desconhecido. Talvez, não volte. Falo, portanto, como quem se confessa" (*Correspondência* 258). [I am two steps from the desert and on the eve of a journey filled with obstacles, the sort that prompts one to compose a will. Perhaps I will not return. I speak, as such, as one who confesses.]

8. In a January 1905 letter to Afonso Arinos, he first describes Manaus as the "ruidosa, ampla, mal-arranjada, monstruoso e opulenta capital dos seringueiros" where fevers rack his body, yet he closes by praising its cosmopolitanism and wondering why "este nosso Brasil" stills exists as periphery on both local and global levels. "Realmente," he muses, "cada vez mais me convenço que esta deplorável rua do Ouvidor é o pior prisma por onde toda a gente vê a nossa terra" (*Correspondência* 251).

9. In a January 1905 letter to Afonso Arinos, he first describes Manaus as the "ruidosa, ampla, mal-arranjada, monstruoso e opulenta capital dos seringueiros" where fevers rack his body, yet he closes by praising its cosmopolitanism and wondering why "este nosso Brasil" stills exists as periphery on both local and global levels. "Realmente," he muses, "cada vez mais me convenço que esta deplorável rua do Ouvidor é o pior prisma por onde toda a gente vê a nossa terra" (*Correspondência* 251).

10. See my essay "Provincializing World Geography: Land and Letters in Euclides da Cunha's Os Sertões" in *Romance Notes* 52.2 (2012): 113–121.

11. See Susanna B Hecht and Alexander Cockburn's *The Fate of the Forest: Developers, Destroyers, and Defenders of the Amazon* (Chicago: U of Chicago P, 2010 [1990]) for a vivid account of the human populations left to deal with years of environmental degradation.

12. For a compelling illustration of geography's relations with empire, see Markham's *Travels in Peru and India While Superintending the Collection of Chinchona Plants and Seeds in South America and their Introduction into India* (London: J. Murray, 1862).

TWELVE

Down Under

New World Literatures and Ecocriticism

George B. Handley

Postnational yearnings in the Americas have yielded a deeper understanding of the submarine "parallel in histories," to quote from Edouard Glissant, throughout the Caribbean basin and beyond (*Caribbean* 63). African, Asian, and European diasporas, and more recent regional migrations between and among Latin American and Caribbean nations and the United States have challenged the very utility of literary regionalisms that have enjoyed long and fruitful histories, particularly in the U.S. South and in Latin America. It is by now well known that regionalism failed in its national aims, but perhaps a new form of regionalism may still prove useful to us in a more advanced global age, one that is attuned to the diasporic histories of human communities and the ecological contexts in which they have taken place.

Regionalism's emphasis on local color, ostensibly in the service of expressing national character, was paradoxically most often drawn to populations whose racial and cultural roots extended well beyond the limited and delimiting bounds of the nation. Cultural particularity, in other words, was sought after as a sign of a new community. Patrician writers and critics paraded the representations of the local as evidence of the long-sought-after rootedness of New World communities, transplanted since 1492 by the globalizing forces of western modernity, and now standing proudly independent, distinct, or even exceptional. Cowboys and Indians, farmers from Scandinavia, former slaves, gauchos of the pampa, guajiros of the monte, mestizos wandering the llanos, all seemed to satisfy the thirst for ample and documentable evidence of exceptional nationalities. However, just as the oral traditions of a remote Amazonian tribe seem to receive their death notice once recorded by a well-intentioned and self-denying western anthropologist, local color now reads more like the last paroxysms of alterity, like so many flightless wings in the entomologist's butterfly collection. Stopping short of tracing the roots of the local back across routes of interdependent complexity and cross-cultural geographical reach, literary regionalists often gave in to the impulse of mythology and fixed the "folk" within spatial territories of stable frontiers and shallow time. Community, in other words, was only loosely geographical and loosely historical in the end, since it upheld a notion of a place that was fixed and static and containable and a sense of history that was far too shallow. Regionalism was, in other words, doomed from the beginning because it offered a temporal freeze frame of diasporic histories and intermixing peoples who were the result of both forced and creative changes and because it offered a spatially frozen frame as well. National boundaries cut through and divided what were otherwise fluid zones of ecological exchange that were most often denied their central relevance to the formation of culture.

Correcting this appears to be a matter of being more historically responsible and chasing with undying determination the fruit at the end of every branch on the family tree. Such has been the conviction of many writers of the postnational global age, invoking the power of fiction to imagine the genealogical reach of kinship stretched across geopolitical borders. Unlike old-time regionalists, these writers understand that not all family trees are created equal since they are not all reconstructed with the same biological honesty to account for what Michel Foucault calls "the accidents, the minute deviations—or conversely, the complete reversals—errors, the false appraisals, and the faulty calculations that gave birth to those things that continue to exist and have value for us" (146). This kind of search, although interested in the meanderings of genealogy, paradoxically "opposes itself to the search for origins," because it understands that the present is not the logical consequence of events and intentions but is constructed by arbitrary accidents (140).

Genealogy, then, works in the hands of these writers as a counterweight to political machinations that divide and conquer the familiar community. As Celia, in Cristina García's *Dreaming in Cuban* (1992), complains presciently prior to the exacerbations of the Cuban Revolution and the U.S. embargo:

> If I was born to live on an island, then I'm grateful for one thing: that the tides rearrange the borders. At least I have the illusion of change, of possibility. To be locked within boundaries plotted by priests and politicians would be the only thing more intolerable.

> Don't you see how they're carving up the world, Gustavo? How they're stealing our geography? Our fates? The arbitrary is no longer in our hands. (99)

Celia's geography is both literal and metaphorical; it signifies the physical space of geopolitical borders and the imaginative space of family bonds across those borders. When the freedom to move physically is denied, the recompense is imaginative freedom, the freedom that sustains the fictional world Garcia creates; "Going South," to borrow a title of one of García's chapters, is to engage in the transgressive ways of knowing—dreaming, singing, dancing, invoking the loas, writing—that enable the women in the novel to see their kinship despite the obstructions imposed by patriarchy. To go south is to imagine a diasporic community—which is what it means to dream in "Cuban" for García—so as to elude the neo-patrician force of such "carvings" that continues to hold national borders inviolate, despite increasing evidence the novel exposes of the Americas' parallels. Cuba, then, is not a nation based on a common language and the imagined community that emerges from it, but rather the imaginative space of longing that connects related geographical spaces and imagined communities that have been severed from one another by political violence. It is a new language of the imagination that exceeds the limits of nation languages. Indeed, in a Global South, genealogy can no longer function as the narrative of what Glissant calls "root identity" and its concomitant "possession of a land," similar to the impulse of much literary regionalism (*Poetics* 143); rather than tapping into a past envisioned as the moment of the world's creation, genealogy becomes an expression of kinship, of close relations ("close" holds "relations" in a paradox; they are admittedly "almost" relations at the same time that they are declared in the family) that effectively expose the rhizomatic family tales of diasporic wanderings, "route" identities that understand themselves relationally with other lands and peoples across space rather than back through time. In other words, a postnational and postcolonial search for roots marks a move away from history, from a temporal axis, to geography and a spatial axis; it brings ecology into view.

For such counter-memories to prove effective in resisting the effects of globalization, however, a Global South must avoid the temptation to want to hide or downplay the rhetorical and contingent nature of such counter-memories. Genealogy, even when it is traced diasporically, holds the allure of an alter-native home, a mythological elsewhere that disrupts the New World subject's sense of place. As I have been implying, naïve regionalisms of New World nativity have never been interested in assessing the reality of transplantation and amnesia that characterize so much New World history, but neither are postnational neo-native appeals to a prior land of origin. Indeed, what is often created by such fiction, a risk

García herself runs, is the impression of a sentimental recovery of a lost home, a new transnational wholeness that simply displaces the merely local with a regionalism more broad in its geographic scope but no less impatient with the New World's lack of historical depth, compared to Europe, and no less interested in genealogical legitimacy and authenticity. The cost of such nostalgia, as Derek Walcott has argued, is the New World itself, its landscapes, seasides, watersheds, and skies that instead of providing inspiration for a newly placed imagination are allowed to degrade because their rawness is a reminder of homelessness and an accusation of being without history, without legitimacy.[1] That is, historical knowledge of the pathways of diaspora will undoubtedly always be useful, but investigations into New World pasts must also reckon with the empty spaces of amnesia that the colonial violence of the hemisphere has wrought.[2] While the imagination of a Global South wants to resist the limitations imposed by the colonial machine, it must also resist the seduction of the appearance of its own limitlessness; otherwise it cannot assess the contours of the violence which rendered such imaginative transgressions necessary. If the ordinary stuff of nature that violent history leaves behind cannot satisfy the more historically impatient writers, they might prefer praising family trees to trees themselves. A writer facing the emptiness of the physical environment can easily give in to the age-old tendency in regionalist literature to mythologize land as territory, as dispensable backdrop to the more central unfolding human drama. However, those writers who continue their search for history into the land itself learn to see behind nature's seeming emptiness a profound confusion between culture and nature and a collusion of melancholic amnesia and elate discovery.

The point here is that the imagination of a Global South is not politically naïve, but it is more importantly profoundly ecological, even or especially when it isn't overtly environmental. We don't have to be engaged in the search for the Latin American version of John Muir, in other words Latin American literature, even when it is not overtly and thematically focused on environmental questions is often engaged, at least implicitly, in ecology, and this is at least for four reasons. First, if we go looking for the solitary figure in the woods as our talisman of nature writing, Latin America will disappoint; Latin America's Catholic roots, as opposed to Protestantism, have made it less likely that the implicit theology of transcendentalism and the ideology of rugged individualism would draw as many writers into the individual and mystical encounter with the world. The encounter with the natural world is a more collective enterprise and is less often represented in overtly edenic language. This was already evident in the differences between Spanish and English Romanticism. Second, Latin America's middle class was much slower to emerge; natural resources were largely controlled by the landed oligarchs whose influence and power were pervasive in the independence movements and in the consequent government. The oppression of the majority poor, in other words, was always and already involved in the exploitation of nature. It is for this reason that questions of social justice have been particularly pronounced in Latin American literature and often create the appearance, at least, of a diminished importance for the natural world in the literature. Third, in Latin America the level of exploitation of nature was on a much higher scale of magnitude and its various exploiters exercised power from the metropolitan centers and from remote colonial centers of power in Spain, England, and in the United States, rather than regionally. Arguments for the protection of nature, then, were somewhat buried in the rhetoric of postcolonial independence, and sometimes buried intentionally due to reluctance to celebrate either British or American influences. And finally, let us not forget the differences of a tropical environment which offered places of far less romantic allure, given their inherent unmanageability. Nature presents itself, in other words, as far more hostile to human interests and concerns.

So to read at least Latin American literature ecocritically, we must confront head on, rather than steer around, the region's political and postcolonial questions. Moreover, it becomes necessary to understand the natural fact that human and natural histories are not separate genealogies but are instead ambiguously and perpetually confused, and that natural processes of mutation and regeneration provide at least one important source of arbitrary accidents that disrupt the consequential logic of human history. The human atrocities of New World history and its various colonial legacies make it seem almost an afterthought, and perhaps a banal one at that, to say that nature too has been a victim of the colonial machine, but the banality has more to do with our incapacity to absorb multiple horrors than with some notion of natural history's comparable insignificance. In many cases New World landscapes were so dramatically altered as a result of the transplantations of western expansion that it is almost anachronistic to talk of "native" plants or to conceive for that matter of a natural history independent of human family trees. But this will require more than dragging New World literature to the tribunals of justice in order to expose complicity with such destruction, usually by means of deconstruction of the uses of nature. What is called for is a fundamental reorientation of culture toward the natural world, one that assures that nature is protagonist not scenery. Poets and novelists on the whole have more to teach in this regard than the poststructuralist criticism of the late twentieth century. It is time for critics, especially those concerned with the social fate of the Global South, to move beyond Foucault's merely political and anthropocentric critique of epistemic machinations and into the realm of nature's silent and untranslatable record of human hands.

Some years ago now, José David Saldívar argued that Havana might be a suitable site for the "Other America," a geography of political resistance to the contemporary embodiments of the Western Zeitgeist. Havana's *Vedado* neighborhood was once the neighborhood of Havana's most privileged white Cubans. García's Celia, for example, describes the pre-revolution neighborhood: "the destitute were everywhere. . . . Remember . . . the beggar families from the countryside looking for work in the iron-fenced mansions of Vedado? The smart couples in their convertible roadsters driving by without a second glance?" (98). It is now the site of Cuba's *Casa de las Américas* (which inhabits a former church), a critical center Saldívar cites for cultural production and criticism from the left. The neighborhood itself has come to signify the great triumph of the Cuban Revolution over the class hierarchies that this neighborhood represented.

But this political critique of American Studies neglects the ecological context. Lost in the ironic story of the once destitute (most of whom were of African descent) who now occupy these former mansions is the story of the loss of landscape and ecological health effectuated by the plantation economy, as Antonio Benítez-Rojo has recently written. Sugar brought slavery, to be sure, but it also brought ecological trauma to the entire Caribbean region, arguably the most environmentally altered region in the world. By the 1630s, many of Havana's "nature preserves" were threatened, including the *Monte Vedado* (the Forbidden Woods) the ultimate destruction of which provided the neighborhood its space and name (Benítez-Rojo 38). If we are to take the Vedado story as a symbol, perhaps its symbolism is as much ecological as it is political; perhaps it means that we ought to look to more environmentally oriented centers, like Cuba's *Fundación Antonio Nuñez Jiménez de la Naturaleza y el Hombre,* that will redress our neglect of the physical world as well as social injustice.

This is not the argument of the Deep Ecologists that all life forms are of equal ethical value, an argument that has too frequently failed to redress the plight of the poor and to address the ways in which social and environmental ills often cause and exacerbate each other. Slavery's evils were social *and* environmental, but this is not to say that we must pick our poison. The colonial machine, after all, did not. Deforestation has been devastating to the integrity of ecosystems and human communities alike, especially when the latter, such as the Caribbean's Neo-African religions, are animistic and tree-centered (Paravisini-Gebert). The recent floods in Haiti were caused by heavy rainfall, yes, but even more by the political and economic forces that have perpetuated the extensive deforestation begun by slavery; destitute descendants of slaves are forced to rely on wood for fuel and thus continue to find their hold on their sense of place slipping away. The consequences of natural events are often distributed according to the tragedies of human oppression and poverty, and for precisely that reason, ecology cannot be ignored in a Global South.

Of course, speaking for nature is as problematic as speaking for the subaltern or for the repressed past, and the political exigencies of the global south are enough to discourage many otherwise fine writers from having to admit that they are essentially engaged in the impossible task of translating the untranslatable. To so confess sounds too much like failure or concession, and many choose instead to continue the search for root identities. In such cases, nature remains, as the tidal movements do for Celia, an illusion of change and mere backdrop to the human story but not as the enigmatic and alluring repository of history's motions and hence a needed source of inspiration for the creation of a new politically unbounded yet nature-bound culture.

Pablo Neruda flirted his entire career with the seductions of a merely political critique of the New World, but what always seemed to disrupt his ideological readings of events was his awareness of natural history's deep time and the need for society to learn from its regenerations, even when this challenged the urgency of his own political aims. In his *Canto General*, he attempts a recuperation of histories lost by the workings of colonial violence across the American hemisphere. He began with an interest in his native Chile, but he soon discovered its experiences are diasporic and that "the roots of all Chileans extended beneath the soil and came from other territories" (qtd. in Santí 20). Although he ostensibly wishes to write a totalizing story of human and natural histories, he becomes increasingly aware of nature's elusiveness and thus of his own limitations. He confesses to an

> ambición de una poesía que englobara no sólo al hombre sino a la naturaleza, a las fuerzas escondidas, una poesía epopéyica que se enfrentara con el gran misterio del universo y tambien con las posibilidades del hombre. (qtd. in Santí 25)
>
> [ambition for a poetry that would include [literally "englobe"] not only man but nature and its hidden forces, an epic poetry that would confront the great mystery of the universe as well as the possibilities of man.]

Although his ambition is totalizing in its scope ("englobara"), it is kept in check by the aesthetic inspired by an encounter with "the great mystery of the universe."

The poem pauses to reflect on the poetic and procreative nature of the history he constructs, since, despite his solidarity and compassion for human suffering and natural devastation, he remains ignorant of the full story: "Estoy hecho de tus raíces, / pero no entiendo, no me entrega / la tierra su sabiduría [I am made of your roots, / but I do not understand, the earth / will not concede its wisdom]" (464). All that he knows is what historiography tells him were the actions of the so-called protagonists, and so he pleads to be led down into the very belly of the earth "hasta llegar/ a la boca del oro [until I reach/ the mouth of gold]" where he hopes that the natural oracle will speak the tale of "la desdicha [sorrows]" of New World history (465). But as he discovers in "The Heights of Macchu Picchu," this tale has been eroded by ecological forces and has been carried by the Andean watershed to the sea where it leaves what he calls in his great ecological poem, "The Great Ocean," "una herencia herida/ bajo el mar [a wounded inheritance/ under the sea]" (772).

Neruda's ocean, unlike Celia's, presents more than an illusion of change but is the very context of perpetual regeneration and transformation in which all human suffering takes place. The ocean is the repository of New World dead and thus contains the stories New World colonialism has buried, but precisely because of nature's capacity to regenerate, the stories will forever be transformed into the stuff of the sea. The world-traveler and ardent voice against human injustice could not resist the impulse to observe and make poems about seashells, rocks, and other detritus at his feet near his home. But this was no turning away from his human concerns. Even though his poetry of nature reaches the most sublime heights of praise, it is never politically indifferent. His poetry, in other words, serves as a reminder to other Latin American writers and to critics as well that you do not have to choose between concerns for human justice and appreciation for ecology. This is because New World history requires that the former turns us to the latter and the latter to the former. If nature and culture have already and always been confused by colonial history, if we cannot determine the boundaries of nations or the distinctions between human communities without doing violence to the integrity of interdependent and complex systems that allow such diversity to emerge, then the human compassion that drives much postcolonial criticism must also be attuned to ecology and environmental history. Neruda's poetry is neither nostalgic for other lands or for the past nor offended by nature's seeming indifference to human tales of sorrow; instead it demonstrates that "going south" must eventually mean taking him "down under" to the *bio*regional spaces—submarine and subterranean—where language must concede its power to control knowledge. Literature of nature, then, does not have to oppose the political; it can instead point the way to a poetics that does not deny the confluence of human and natural histories. That this poetics would become politically relevant is, for the imagination of the Global South, just a natural part of the dialectic of recovering a sense of place in the wake of history's wreckage.

Because it always implies interdependence and merely rhetorical boundaries, foregrounding ecology brings political and genre boundaries into question and makes a comparative frame necessary—between genres, epistemologies, and geopolitical borders—in order to approach ecology's extensity and complexity. As I have acknowledged, however, broad comparative frames run their risks, just as narrow nationalist ones do. For example, to argue that the Amazon is not an entity that fits within any one nation's borders, for example, is meant to highlight the dangers of monolithic and static definitions of nature and the contingency of nationalism itself. But it ignores political reality to insist rigidly on the oppositional idea that the Amazon belongs to all of the Americas or to the entire world. Candace Slater's work on representations of the Amazon has exposed the attendant problems of contemporary environmentalism in its desire to globalize and thus save the rainforest from the locals. The mono-mythic and cosmopolitan view of the Amazon as the endangered Eden, as the lung of the world, has been a vital catalyst for environmental thought and action. Indeed, ecologically speaking it is one of the more cross-cultural and vital ecologies in the world. If climate change persists unabated, however, we can expect to see signs of the collapse of the Amazon basin and its rainforest by 2040, resulting in a release of carbon stored in its vegetation and soils that equals 8 percent of the world's stored carbon (Flannery 198). However, the idea of the rainforest as an ecosystem tends to elide the rich overlay of stories, myths, and histories that emerge from local and transnational experience within the watershed. Slater's exposé of that overlay makes a strong case for what she calls a "meeting of the waters," or a comparative and dialectic criticism that moves from within and from without a given geographical space. I endorse this dialectical and comparative approach, but I wish to suggest further that it is made necessary by the very ecology of this watershed. Ecological understanding may serve not only to balance the social and cultural concerns she raises but also to aid us in the formation of a dialectic between the global and the local and between culture and nature that is adequate to contemporary environmental problems.

The implication here is that criticism that seeks to organize the chaos of literature into meaningful and discrete systems of knowledge must, like the prose that attends to the intertwining ambiguities of form in the tropics, remain decentered, interdisciplinary, and, we might even say, baroque in its methodology. The baroque form has been widely recognized as a vital strategy for American and Latin American writers such as Alejo Carpentier, William Faulkner, Carlos Fuentes, José Lezama Lima, and others who have sought to represent the complex overlay of histories in the Americas. As I have argued elsewhere, the aesthetics of the New World Baroque is a productive response to ecology; it stipulates the priority of the land, the almost moral obligation of mimesis, and the inevitability of self-reflexive excess in which ecology decenters the human self (Handley, "New World Baroque"). Much Latin American and Caribbean fiction places front and center the complexity, interdependency, and unpredictability of ecosystems; as a result, this literature is similarly resistant to reductive or isolated readings. It demands a recognition of the mutually constitutive forces of nature and culture and the need for shifting and comparing geographical and literary contexts in order to tease out its fullest range of meanings. Criticism, in other words, should not seek to reduce literature, like a dam to a river, to an ideologically fixed point. In comparative and ecological readings, major nature writing from the North becomes less the suspicious enemy and more a valuable point of comparative reflection on similar questions. Such approaches also inspire new readings of American canonical figures, such as Henry David Thoreau. In his description of Thoreau's attempt at biocentric prose, Aaron Sachs pinpoints the reasons why such a decentered response is necessary to respond to the empirical realities of ecology, realities he notes Alexander von Humboldt was making apparent in his Latin American research. Sachs describes how Thoreau allows ecology to deconstruct his human-centered view only to "then buil[d] the whole back up into something even more complex, rich, overwhelming. To be slightly off balance oneself was the only way to make a cosmos of the chaos" (97). When we juxtapose writers in the Americas and foreground their wrestle with giving expression to place in the most profoundly ecological way possible, we begin to see literature's role as more than a creative force behind a geo-political imagined community. Instead we see that language is undone by its own ambition to give expression to the physical world which cultures seek to inhabit. We see imagined communities of human beings in a more-than-human context and we begin to read literatures and cultures as themselves interdependent and contingent ecosystems.

ACKNOWLEDGMENT

This article is reprinted with permission from the editors of the journal *Global South*, where it appeared in issue 1.1 (Winter 2007): 91-97.

WORKS CITED

Benítez-Rojo, Antonio. "Sugar and the Environment in Cuba" *Caribbean Literature and the Environment*. Ed. Elizabeth M. DeLoughrey, Renée K. Gosson, and George B. Handley. Charlottesville: U of Virginia P. 2005. 33–50. Print.

Foucault, Michel. *Language, Counter-memory, Practice: Selected Essays and Interviews by Michel Foucault*. Ed. Donald F. Bouchard. Ithaca, NY: Cornell UP, 1977. 139–64. Print.

Flannery, Timothy. *The History and Future Impact of Climate Change*. London: Allen Lane, 2006. Print.

García, Cristina. *Dreaming in Cuban*. New York: Ballantine, 1992. Print.

Glissant, Edouard. *Caribbean Discourse; Selected Essays*. Charlottesville: U of Virginia P, 1989. Print.

———. *Poetics of Relation*. Trans. Betsy Wing. Ann Arbor: U of Michigan P, 1997. Print.

Handley, George B. "New World Baroque as Postcolonial Ecology." In *Postcolonial Ecologies: Literatures of the Environment*. Ed. Elizabeth DeLoughrey and George Handley. Oxford, UK: Oxford UP, 2011. 117–35. Print.

———. *New World Poetics: Nature and the Adamic Imaginaton of Whitman, Neruda, and Walcott*. Athens: U of Georgia P, 2010. Print.

Neruda, Pablo. *Canto General*. Ed. Enrico Mario Santí. Madrid: Cátedra (LetrasHispánicas), 2000. Print.

Paravisini-Gebert, Lizabeth. "'He of the Trees': Nature, Environment, and Creole Religiosities in Caribbean Literature." In *Caribbean Literature and the Environment*. Ed. Elizabeth M. DeLoughrey, Renée K. Gosson, and George B. Handley. Charlottesville: U of Virginia P, 2005. 182–96. Print.

Sachs, Aaron. *The Humboldt Current: Nineteenth-Century Exploration and the Roots of American Environmentalism*. New York: Penguin, 2007. Print.

Saldívar, José David. *The Dialectics of Our America: Genealogy, Cultural Critique, and Literary History*. Durham, NC: Duke UP, 1991. Print.

Slater, Candace. *Entangled Edens: Visions of the Amazon*. Berkeley: U of California P, 2002. Print.

Walcott, Derek. "The Muse of History." In *What the Twilight Says*. New York: Farrar, Straus, and Giroux, 1998. Print.

NOTES

1. See his "Muse of History."
2. For elaboration on this point, see my book, *New World Poetics*, especially chapter two.

Index

About the Contributors

George B. Handley is Professor of Interdisciplinary Humanities and Chair of the Department of Humanities, Classics, and Comparative Literature at Brigham Young University. His books include *New World Poetics: Nature and the Adamic Imagination of Whitman, Neruda, and Walcott* and the co-edited volumes *Postcolonial Ecologies: Literatures of the Environment* and *Caribbean Literature and the Environment: Between Nature and Culture.* His articles have appeared in such journals as *Callaloo, Modern Fiction Studies, ISLE, Mississippi Quarterly,* and others.

Cheng Li is an M.A. student in the Department of English at Beijing International Studies University in the People's Republic of China. His research fields include ecocriticism, environmental literature, and comparative literature.

Yanjun Liu is a Ph.D. student in the Department of Political Science at the University of California, Santa Barbara. He is interested in environmental politics, political participation, and the political economy of development.

Aarti S. Madan completed her Ph.D. at the University of Pittsburgh and is an Assistant Professor of Spanish and International Studies in the Department of Humanities & Arts at Worcester Polytechnic Institute, where she directs the Buenos Aires Language & Culture Immersion Program. Her primary research interests include the intersection between literature and geography in Latin American narrative, particularly during the period of nation-state consolidation. Her newest project explores literary and cultural exchanges between the Iberoamerican and Indian spheres. Most recently she has published articles in *MLN, Hipertexto, Romance Notes,* and *Dissidences.*

Salma Monani is an Assistant Professor in Gettysburg College's Environmental Studies Department. She is co-editor of *Ecocinema Theory and Practice.* She has also published her research, which includes explorations of film and environmental justice, film festival studies, and indigenous eco-activism, in journals such as *ISLE, The Journal of Nature and Culture,* and *Local Environment,* and in various anthologies.

Pramod K. Nayar teaches in the Deptartment of English at The University of Hyderabad in India. His most recent books include *Frantz Fanon, Posthumanism, Digital Cool, Colonial Voices: The Discourses of Empire,* and *Writing Wrongs: The Cultural Construction of Human Rights in India,* besides essays on posthumanism (*Modern Fiction Studies*), Indian travel writing (*New Zealand Journal of Asian Studies, South Asian Review*), and graphic novels (*Hungarian Journal of American and English Studies, South Asian Film and Media*). He is currently working on the Postcolonial Studies Dictionary, a book on transnationalism in English literature, and a book on surveillance. He is often seen reading superhero comics in his office.

Dora Ramirez-Dhoore is an Associate Professor of American literature at Boise State University. Her research examines the production and consumption of texts tied to global and transnational perspectives of audience. Her work also incorporates ideas of nation-building while examining the internalization of socio-political global affects within the Latina/o population. Her scholarly work has been published in *Feminist Teacher, Reflections: A Journal of Writing, Service-Learning, Community Literacy, Chicana/Latina Studies, The Natural World in Latin American Literatures: Ecocritical Essays on Twentieth Century Writings,* and other journals and collections. She holds a Ph.D. in Literature from the University of Nebraska-Lincoln.

Swarnalatha Rangarajan has been Associate Professor of English at the Department of Humanities and Social Sciences at the Indian Institute of Technology Madras since 2010. She was a Fulbright Pre-doctoral fellow at Harvard University in 1999–2000. She was the founding editor of the *Indian Journal of Ecocriticism* (IJE) and has served as Guest Editor for two special issues on Indian ecosophy for *The Trumpeter*—the Canadian Journal of Deep Ecology. She has published articles on ecocriticism in several journals. She has been recently awarded the Charles Wallace Fellowship at CRASSH (Cambridge University) to pursue a book project on ecocriticism. Her short fiction has appeared in anthologies from publishing houses like Penguin, Zubaan, and Westland, to name a few. She has also completed her first novel, which has a prominent ecosophical theme.

Vidya Sarveswaran works as an Assistant Professor of English in the Department of Humanities and Social Sciences at the Indian Institute of Technology Jodhpur. She was a Fulbright Fellow for the year 2008–2009 at the University of Nevada, Reno. She completed her undergraduate and M.A. programs at the University of Madras and her doctoral degree at IIT Madras. She was the student editor of the *Indian Journal of Ecocriticism,* has several journal articles to her credit, and has worked on several documentary films as a script writer. She is interested in creative writing, film studies, and environmental humanities.

Scott Slovic served as founding president of the Association for the Study of Literature and Environment (ASLE) from 1992 to 1995 and has edited the journal *ISLE: Interdisciplinary Studies in Literature and Environment* since 1995. The author of many books and articles in the field, he is currently writing *Fundamentals of Ecocriticism and Environmental Literature* and editing *The Cambridge Companion to American Literature and Environment*, among other projects. He is professor of literature and environment at the University of Idaho.

Inna Sukhenko is an Associate Professor in the Mass Media Studies Department of the Mass Media and Communication Faculty at Dnipropetrovsk National University in Ukraine. She graduated from Dnipropetrovsk National University in 1997 and received her doctoral degree in 2003. She is a grantee of the Erasmus Mundus External Cooperation Window Mobility Scheme (Bologna, Italy, 2008), Erasmus Mundus Action 2 BMU-MID (Belarus, Moldova, Ukraine-Mobilities for Innovations and Development) (Turku, Finland, 2011), and the Cambridge Colleges' Hospitality Scheme (Cambridge, UK, 2013). She is interested in literary studies, nature writing, and cross-cultural communication.

Tsutomu Takahashi is Professor of American Literature in the Faculty of Languages and Cultures at Kyushu University, Japan. He earned his Ph.D. in Comparative Literature at Pennsylvania State University. His recent works include *Concord Jeremiad: Thoreau's Rhetoric of the Age* (in Japanese) and the co-edited volume *Atlantic Imagination: Crossing-Over of American Renaissance* (in Japanese).

Karen Thornber is Professor and Chair of the Department of Comparative Literature at Harvard University. Her research and teaching focus on world literature and the literatures and cultures of East Asia. She is the author of two international prize-winning books: *Empire of Texts in Motion: Chinese, Korean, and Taiwanese Transculturations of Japanese Literature* and *Ecoambiguity: East Asian Literatures and Environmental Crises*. These books draw on extensive archival research in East Asia. Thornber is currently writing two books, one on world literature and global health, and the other on world literature, gender, and violence, for which she is studying Hindi, Urdu, and Arabic.

Jyotirmaya Tripathy is in the Faculty of Humanities and Social Sciences at the Indian Institute of Technology Madras. His areas of interest are cultural studies and critical development studies. He has been making interventions in these areas in journals like *Development in Practice, Journal of Developing Societies, International Journal of Cultural Studies, Journal of*

Third World Studies, and others. He is also the co-editor of *The Democratic Predicament: Cultural Diversity in Europe and India*.

Laura A. White is Assistant Professor of Anglophone Literature at Middle Tennessee State University, where she is also affiliate faculty with the Women's and Gender Studies and African Studies programs. Building on intersections between postcolonial ecocriticism and material ecocriticism, her current research investigates ways that narrative strategies operate to position the human in relation to nonhuman matter in contemporary Australian, South Asian, and African novels. Recent publications include essays in the *Journal of Commonwealth Literature, ISLE*, and the edited collection *International Perspectives in Feminist Ecocriticism*.

Gang Yue is Associate Professor in the Department of Asian Studies at the University of North Carolina at Chapel Hill. He is the author of *The Mouth That Begs: Hunger, Cannibalism, and the Politics of Eating in Modern China*. His present research is concerned with modern Tibet and the rise of China as related to development and the environment.